ENGINEERING OF
CHEMICAL COMPLEXITY II

WORLD SCIENTIFIC LECTURE NOTES IN COMPLEX SYSTEMS

AIMS AND SCOPE

The aim of this new interdisciplinary series is to promote the exchange of information between scientists working in different fields, who are involved in the study of complex systems, and to foster education and training of young scientists entering this rapidly developing research area.

The scope of the series is broad and will include: Statistical physics of large non-equilibrium systems; problems of nonlinear pattern formation in chemistry; complex organization of intracellular processes and biochemical networks of a living cell; various aspects of cell-to-cell communication; behaviour of bacterial colonies; neural networks; functioning and organization of animal populations and large ecological systems; modeling complex social phenomena; applications of statistical mechanics to studies of economics and financial markets; multi-agent robotics and collective intelligence; the emergence and evolution of large-scale communication networks; general mathematical studies of complex cooperative behaviour in large systems.

World Scientific Lecture Notes
in Complex Systems – Vol. 12

editors

Alexander S Mikhailov
Gerhard Ertl

Fritz Haber Institute of the Max Planck Society, Germany

ENGINEERING OF
CHEMICAL COMPLEXITY II

 **World Scientific**

NEW JERSEY · LONDON · SINGAPORE · BEIJING · SHANGHAI · HONG KONG · TAIPEI · CHENNAI

Published by

World Scientific Publishing Co. Pte. Ltd.

5 Toh Tuck Link, Singapore 596224

USA office: 27 Warren Street, Suite 401-402, Hackensack, NJ 07601

UK office: 57 Shelton Street, Covent Garden, London WC2H 9HE

British Library Cataloguing-in-Publication Data
A catalogue record for this book is available from the British Library.

World Scientific Lecture Notes in Complex Systems — Vol. 12
ENGINEERING OF CHEMICAL COMPLEXITY II

Copyright © 2015 by World Scientific Publishing Co. Pte. Ltd.

ISBN 978-981-4616-12-6

In-house Editor: Christopher Teo

Printed in Singapore

PREFACE

With this second volume we continue publication of surveys in a novel and rapidly developing field whose subject is the understanding, design and control of self-organized chemical systems. The previous volume appeared at World Scientific in 2012 and presented plenary and selected invited talks at the international conference *Engineering of Chemical Complexity* organized in 2011 by the Berlin Center for Studies of Complex Chemical Systems. The present book is based on the subsequent conference with the same title which has taken place in the summer of 2013 in the beautiful seaside resort of Warnemünde on the Baltic coast of Germany. The authors were chosen by us among the invited speakers; we have asked them to provide brief reviews of those promising research directions in which they currently work. The volume covers a broad range of topics, from molecular machines and problems in cell biology to chemical oscillations and waves in macroscopic reactors. It is fascinating to see how the research is steadily spreading to new areas and how fundamental engineering applications, such as synthetic molecular motors, specially designed active matter and reactive gels, are developing and gradually implemented. We would like to acknowledge valuable support in the organization of the Warnemünde conference by Ulrike Christine Künkel and Shigefumi Hata. Our special thanks go to Maximillian Eisbach who has assisted us much both in the organization of the conference and in the preparation of this volume.

Alexander S. Mikhailov and Gerhard Ertl

CONTENTS

Chapter 1

FROM SIMPLE TO COMPLEX OSCILLATORY BEHAVIOR IN CELLULAR REGULATORY NETWORKS

Albert Goldbeter[*,†] and Claude Gérard[*,‡]

*Unité de Chronobiologie théorique, Faculté des Sciences, Université Libre de Bruxelles (ULB), Campus Plaine, CP 231, B-1050 Brussels, Belgium

†Stellenbosch Institute for Advanced Study (STIAS), Wallenberg Research Centre at Stellenbosch University, Marais Street, Stellenbosch 7600, South Africa

‡Current address: de Duve Institute, Université Catholique de Louvain (UCL), Avenue Hippocrate 75, B-1200 Brussels, Belgium

1. Introduction

Oscillatory behavior is observed at all levels of biological organization, with periods ranging from a fraction of a second to years.[1–3] The fastest rhythms occur in neurons and muscle cells. Neuronal oscillatory networks control major physiological functions such as respiration.[4] The brain in fact generates a huge variety of rhythms with widely different frequencies,[5] while cardiac rhythmicity is an autonomous property of nodal tissues of the heart.[6] Rhythms of non-electrical nature nevertheless abound in cellular systems. Thus, sustained oscillations are observed in the glycolytic pathway in yeast, with a period of several minutes, as a result of enzyme regulation.[7–9] Periodic variation of intracellular calcium, with a period ranging from seconds to minutes, is observed in many cell types upon stimulation by a hormone or a neurotransmitter.[10,11] Oscillations and waves of cyclic AMP (cAMP) control the periodic aggregation of *Dictyostelium* amoebae after starvation.[12] The segmentation clock controls the periodic formation of somites in embryogenesis.[13] A network of enzyme reactions that behaves as a self-sustained oscillator drives the cell cycle.[14,15] Lastly, endogenous circadian rhythms of a period close to 24 h occur in eukaryotic organisms and some prokaryotes like cyanobacteria.[16–18] This list is by no means exhaustive, and several new examples of oscillatory behavior at the cellular level have been uncovered in recent years.[19]

Why are there so many biological rhythms? Biological systems are prone to oscillate because they satisfy three conditions: they are open as they exchange matter and energy with their environment, they generally operate far from equilibrium, and at all levels of biological organization, subcellular, cellular or supracellular, they are governed by nonlinear kinetic laws.[1] Such nonlinearities create the conditions for instability: it is precisely when a non-equilibrium steady state becomes unstable that sustained oscillations develop, especially when the evolution equations admit a single steady-state solution. Such oscillations are particularly robust since they correspond to the evolution to a limit cycle in phase space, so that their amplitude and frequency do not depend on initial conditions. Cellular rhythms represent a process of non-equilibrium self-organization and can therefore be viewed as temporal dissipative structures.[20]

What are the main sources of nonlinearity needed for eliciting the instabilities that lead to oscillatory behavior? A major source is regulatory feedback. Feedback regulation, through activation or inhibition, can take many forms, depending on the level at which it is exerted. Thus, if we focus on cellular processes, positive or negative feedback may be exerted on the expression of genes, on the activity of enzymes or receptors, on transport between different organelles and the intracellular milieu, or on the conductance of ionic channels, to name but the major types of cellular control. Each, and sometimes several, of these modes of regulation are involved in known examples of cellular rhythms. Thus, enzyme regulation underlies glycolytic oscillations, transport regulation is involved in the periodic evolution of intracellular Ca^{++}, the regulation of ion channels is at the core of oscillations of the membrane potential in nerve and muscle cells, while circadian rhythms originate from the control of gene expression.[1]

Oscillations do not take only the form of simple periodic behavior. More complex oscillatory phenomena can also occur in regulated cellular networks, as in other fields of chemistry and physics. Thus, complex periodic oscillations in which successive trains of high-frequency spikes are separated by phases of quiescence (bursting) are observed in many nerve cells such as the R15 neuron in Aplysia.[21] Aperiodic oscillations in the form of chaos can also occur in some systems, often as a result of the periodic forcing of a limit cycle oscillator.[22] Mathematical models and numerical simulations are useful in studying the conditions in which simple and complex oscillatory phenomena occur in regulated cellular networks.[23] Thus, models bring to light the role of multiple instability-generating mechanisms in extending the repertoire of oscillatory dynamics from simple periodic oscillations to more complex modes of oscillatory behavior. Here we use the modeling approach to examine the transition from simple to complex oscillatory behavior. First, we consider a relatively simple biochemical model, which has been used for long as a prototype in studying the transition from simple to complex oscillatory behavior. Then we turn to a realistic model for the enzymatic network driving the mammalian cell cycle. Given the complexity of its regulatory structure, multiple sources of oscillations exist in this network. As a result, oscillations can take a simple or complex form, including chaos, depending on whether the different oscillatory circuits present within the cell cycle network are tightly coupled or not. We conclude by comparing these results with the mechanisms underlying the transition from simple to complex oscillations in other known examples of oscillatory cellular networks.

2. From Simple to Complex Oscillatory Dynamics in a Prototype Biochemical Model

It is useful to recall here the properties of a simple biochemical model that was built three decades ago to investigate the dynamical consequences of an interaction between two instability-generating mechanisms.[24] This model is based on the coupling in series of two allosteric enzymes, E_1 and E_2 (see Fig. 1).

The first enzyme, E_1, transforms a substrate S into product P_1, while the latter serves as substrate for enzyme E_2, which transforms it into product P_2. Substrate S is injected at a constant rate v, whereas the end product P_2 is removed at a rate proportional to its concentration, characterized by the rate constant k_s. The regulatory structure of this simple network involves two positive feedback loops coupled in series: product P_1 activates enzyme E_1, while product P_2 activates enzyme E_2.

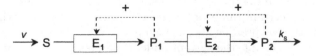

Fig. 1.   Model of two autocatalytic enzyme reactions coupled in series. This model provides a three-variable prototype for studying the transition from simple to complex oscillatory behavior, including birhythmicity, bursting and chaos.[1,24]

Previous work on the role of phosphofructokinase in the mechanism of glycolytic oscillations established that the activation of an allosteric enzyme by its reaction product can produce an instability.[1,25] The allosteric nature of the enzyme is reflected by the sigmoidal form of the reaction rate as a function of the substrate or product concentration and ensures that the reaction rate is sufficiently nonlinear. A model for a single allosteric enzyme activated by its reaction product shows that the evolution equations admit a single steady-state solution that can become unstable in a domain bounded by two critical values of the substrate injection rate. In this domain, the system undergoes sustained oscillations that correspond to a stable limit cycle in phase space.[1,25]

Coupling two positive feedback loops in series, as in the model schematized in Fig. 1, allows for the interplay of two instability-generating mechanisms. The analysis of this three-variable model indicates that the repertoire of dynamic behavior is much enriched: besides the evolution to a stable steady state or to sustained oscillations of the limit cycle type corresponding to simple periodic behavior, the system is capable of displaying (i) complex periodic oscillations, i.e. bursting, corresponding to a folded limit cycle; (ii) aperiodic oscillations, i.e. chaos, corresponding to the evolution to a strange attractor in the phase space; (iii) a co-existence between two simultaneously stable oscillatory regimes, periodic or chaotic such a phenomenon was referred to as birhythmicity.[24]

Further investigation of the model showed that these complex modes of oscillatory behavior occur when the two instability-generating mechanisms are active simultaneously. Chaos occurs, for example, when the oscillation due to the first reaction feeds into the second oscillatory enzyme reaction. This situation becomes analogous to the occurrence of chaotic behavior in an oscillatory system forced by a periodic input. The difference, however, is that chaos is non-autonomous in the latter case, and autonomous in the bi-enzymatic system schematized in Fig. 1 where the two sources of oscillations are inherent to the system.

The fact that the prototype model based on two autocatalytic enzyme reactions coupled in series contains only three variables allows a detailed analysis of the onset of bursting.[26] Thus, by treating substrate S as a slowly varying parameter, bifurcations diagrams of the P_1-P_2 subsystem can be constructed. These diagrams show bistability: two stable steady states of P_1 coexist in a certain range of S concentrations as a result of the positive feedback exerted by product P_1 on enzyme E_1. When the (slow) variation of S is taken into account, the system undergoes repeated cycles of hysteresis along the bistable curve. This behavior corresponds to simple periodic oscillations of the limit cycle type. How does bursting occur in the system?

The second positive feedback loop exerted by product P_2 on enzyme E_2 is capable of inducing an instability of the upper steady state in the range of S values producing bistability. Then, on the higher branch of steady states of P_1 the dynamics switches from a stable to an unstable focus, as a result of a Hopf bifurcation. It is in these conditions that simple periodic oscillations transform into complex periodic oscillations of the bursting type.[26]

Other modes of complex oscillatory behavior can also occur in the system, such as the coexistence between two (birhythmicity) or even three (trirhythmicity) stable periodic regimes.[1] In parameter space, however, the domain of simple periodic oscillations remains larger than that of bursting, which domain is itself larger than those of birhythmicity or chaos. The analysis in a two-variable phase plane can be further simplified by resorting to a one-dimensional return map capturing the dynamical properties of the model.[1,26] Such approach provides a powerful tool for analyzing the onset of bursting and chaos in this system.

The coupling of multiple sources of oscillations, considered in the simple, prototype network of Fig. 1, is found in many known cellular networks in which it is likely to underlie the occurrence of complex oscillatory behavior. In the following we focus on the regulatory network that drives the mammalian cell cycle. This enzymatic network operates as a self-sustained oscillator. We will examine how the coupling between multiple sources of oscillations can produce complex patterns of oscillatory behavior in this key cellular network.

3. Simple Periodic Behavior in the Cdk Regulatory Network Driving the Mammalian Cell Cycle

The mammalian cell cycle is composed of four different phases: G1, S (DNA replication), G2, and M (mitosis). A network of enzymes known as cyclin-dependent kinases (Cdks) controls the ordered progression through the successive phases of the cell cycle: the cyclin D/Cdk4-6 and cyclin E/Cdk2 complexes promote progression in G1 and allow the G1/S transition, the cyclin A/Cdk2 complex ensures progression in S and elicits the S/G2 transition, while the cyclin B/Cdk1 complex brings about the G2/M transition.[14] Cdk regulation is achieved through a variety of mechanisms that include association with cyclins and protein inhibitors, phosphorylation-dephosphorylation, and cyclin synthesis or degradation.[14]

We previously proposed a detailed model for the dynamics of the Cdk network that drives the mammalian cell cycle and explored the conditions for its temporal self-organization.[15,27] The model for the Cdk network

A. Goldbeter & C. Gérard

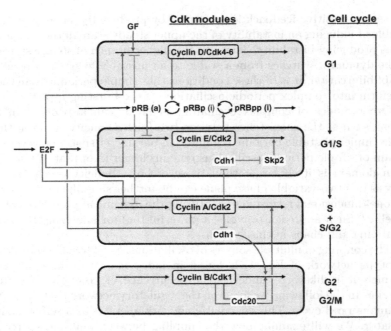

Fig. 2.   Scheme of the model for the Cdk network driving the mammalian cell cycle.[15] The model incorporates four modules centered on the main cyclin/Cdk complexes: cyclin D/Cdk4-6, cyclin E/Cdk2, cyclin A/Cdk2 and cyclin B/Cdk1. Also considered are the effect of the growth factor GF and the role of the pRB/E2F pathway, which controls cell cycle progression (see text and Ref. 15 for more detail).

(Fig. 2) contains four coupled modules centered on cyclin D/Cdk4-6, cyclin E/Cdk2, cyclin A/Cdk2, and cyclin B/Cdk1, respectively (for more detailed schemes of the four modules, see supporting information in Ref. 15). The activity of the cyclin/Cdk complexes is regulated both through phosphorylation (by kinase Wee1) and dephosphorylation (by phosphatase Cdc25), and reversible association with the protein inhibitors p21 or p27. The model includes the Retinoblastoma protein pRB and the transcription factor E2F, which respectively inhibit and promote progression in the cell cycle. The Cdk network itself controls the balance between the antagonistic effects of pRB and E2F through phosphorylation. Additional regulations of cyclin/Cdk complexes occur in the form of negative feedback, which arises from Cdk-induced cyclin degradation, and positive feedback, which originates from the fact that Cdk1 and Cdk2 activate their phosphatases Cdc25 and thereby promote their own activation. The temporal evolution of the model is governed by a set of 39 kinetic equations, which are listed in Sec. 2 in the supporting information of Ref. 15. These equations are based on mass action or Michaelian kinetics. Five additional equations are needed to incorporate into the model a checkpoint on DNA replication. To limit the complexity of the model we only considered the variation of protein levels without incorporating explicitly changes in the mRNAs. For a detailed description of the four modules of the model and their regulatory interactions, see supporting information in Ref. 15 where variables are defined in Table S1, while a definition of parameters and a list of their numerical values are given in Table S2.

We characterized the dynamics of the Cdk network by means of bifurcation diagrams established with respect to the concentration of growth factor GF taken as control parameter.[15] When GF exceeds a critical value, the system crosses a Hopf bifurcation point and repetitive activation of the cyclin/Cdk complexes occurs in the form of self-sustained oscillations (Fig. 3A). The ordered activation of the cyclin/Cdk complexes controls the passage through the successive phases of the cell cycle: the peaks in the activity of cyclin D/Cdk4-6 and cyclin E/Cdk2 allow progression from G1 to S, the activity of cyclin A/Cdk2 rises in S and G2, while at the end of the cycle, the peak in cyclin B/Cdk1 brings about the G2/M transition. The level of cyclin D/Cdk4-6 remains elevated throughout the cycle, in agreement with experimental observations, and falls only when GF is removed. Sustained oscillations shown in Fig. 3A correspond to the evolution, in the phase space, to a stable limit cycle, which can be reached regardless of initial conditions as illustrated in Fig. 3B.

The model accounts for key properties of the mammalian cell cycle such as the need for a fine-tuned balance between pRB and E2F for oscillations to occur, and the existence of a restriction point in G1, which is a point of no return beyond which they are irreversibly engaged in the cell cycle and do not require the presence of growth factor to complete mitosis. When the DNA replication checkpoint is incorporated into the model, this checkpoint was found to slow down the dynamics of the network without modifying its oscillatory nature. Below the threshold level of GF cells are in a stable steady state, which can be associated with the quiescent phase, G0. Above the threshold the repetitive activation of the Cdks can be associated with cell proliferation. As illustrated by stability diagrams established, for example, as a function of cyclins D and E, and as a function of the activity of the phosphatase Cdc25 that activates cyclin A/Cdk2 and of the level of the protein Cdh1 that controls cyclin B degradation, oscillations occur in large domains in parameter space and therefore represent a generic property of the Cdk network resulting from its regulatory wiring.[15]

(A)

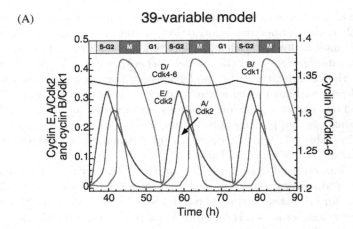

(B)

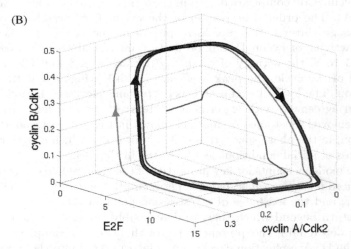

Fig. 3. Simple periodic oscillations of the limit cycle type in the Cdk network.[33] (A) Time series: in the presence of a suprathreshold amount of GF, sustained oscillations correspond to the repetitive, ordered activation of the four cyclin/Cdk complexes. The curves were generated by numerical integration of the kinetic equations (1)–(39) listed in Sec. 2 in the supporting information of Ref. 15, for the parameter values listed in Table S2 (http://www.pnas.org/content/106/51/21643/suppl/DCSupplemental). Shown are the oscillations in the active forms of the cyclin/Cdk complexes. (B) The oscillations in (A) correspond to the evolution to a unique limit cycle regardless of initial conditions. The same closed curve is reached for two distinct initial conditions when the trajectory of the 39-variable system is projected onto the three-variable phase space formed by the concentrations of E2F, cyclin A/Cdk2 and cyclin B/Cdk1.[33]

4. Complex Oscillatory Behavior in the Cdk Network

Even though it can globally operate in a periodic manner, the Cdk network nevertheless contains at least four oscillatory circuits, each of which can produce sustained oscillations on its own.[15] When tightly coupled, as occurs in physiological conditions, these circuits cooperate to produce the periodic, ordered activation of the cyclin/Cdk complexes that drive the successive phases of the cell cycle. It is however possible to isolate in the model for the Cdk network four circuits, all based on negative feedback regulation, which are capable of generating sustained oscillations when isolated from the rest of the network (such isolation can readily be achieved in the model by setting the relevant coupling parameters equal to zero). The four circuits, whose oscillatory properties were confirmed by numerical simulations, are schematized in Fig. 4.

All four oscillators contain cyclin A/Cdk2, but only circuits 3 and 4 also contain cyclin B/Cdk1 and can thus be viewed as mitotic oscillators producing a peak in cyclin B/Cdk1. In contrast, the sub-networks 1 and 2

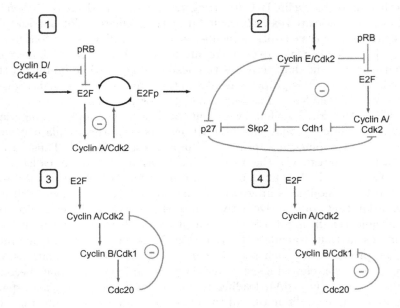

Fig. 4. Multiple oscillatory circuits in the Cdk network. When isolated in the model for the mammalian cell cycle, four circuits containing negative feedback loops are capable of generating sustained oscillations.[15,33] In physiological conditions, all four oscillatory circuits synchronize to produce the ordered, repetitive activation of the different modules driving the successive phases of the mammalian cell cycle.

predict oscillations in cyclin A/Cdk2 in the absence of coupling to Cdk1. The oscillatory circuits 1 and 2 produce Cdk1-independent Cdk2 oscillations and are therefore associated with endoreplication, while circuits 3 and 4 involve Cdk1 oscillations and are associated with periodic cell division.[15] Endoreplication corresponds to multiple passages through the phase of DNA replication without mitosis.[28,29] The possibility of endoreplication was previously reported in a model for the yeast cell cycle[30] and in a generic model for the eukaryotic cell cycle.[31] Rapid cycling likely associated with an oscillatory sub-network involving cyclin B/Cdk1 has been revealed by treatments perturbing the normal operation of the Cdk network.[32]

The sequential activation of the Cdk modules is associated with the temporal self-organization of the Cdk network and with the global, periodic operation of the mammalian cell cycle. The first three modules of the network (see Fig. 2) centered on cyclin D/Cdk4-6, cyclin E/Cdk2, and cyclin A/Cdk2 cooperate to induce the transient firing of the last, embryonic-like, oscillatory module centered on cyclin B/Cdk1. Thus, the first two modules elicit in module 3 the increase in cyclin A/Cdk2, which transiently drives module 4 into the domain of sustained oscillations. The resulting pulse in cyclin B/Cdk1 triggers in turn the decrease in cyclin A/Cdk2, the associated exit of circuit 4 from the oscillatory domain, and the re-instatement of the conditions leading to a new round of oscillations in the cell cycle.[15] The key feature of the regulatory structure of the Cdk network is that each Cdk module activates the next module(s) while inhibiting the preceding one(s). This property results in the temporal self-organization of the network in the form of self-sustained Cdk oscillations.

When the four oscillatory circuits are weakly coupled, numerical simulations indicate that the interactions between the oscillatory sub-networks may sometimes produce complex oscillations.[33] Thus, upon decreasing progressively the rate of inactivation of protein Cdh1 that promotes cyclin B degradation, the rate of cyclin B hydrolysis increases and the Cdk1 module is progressively inactivated. In agreement with the observation that constitutive activation of the protein Cdh1 can uncouple DNA replication from mitosis,[29] such a change leads to the transition from mitotic oscillations to endoreplication (Fig. 5): several peaks in Cdk2 may then be produced for each peak in Cdk1. These periodic oscillations are more or less complex (Figs. 5B and 5C). Sometimes they may become highly complex (Fig. 6A), leading to folded limit cycles in phase space (Fig. 6B). The periodic nature of the trajectory can be ascertained by the construction of a Poincaré section, which shows a limited number of fixed points (Fig. 6C).

Chaotic oscillations (Fig. 7A) can also occur for parameter values close to those producing quasiperiodic or complex periodic behavior. In phase

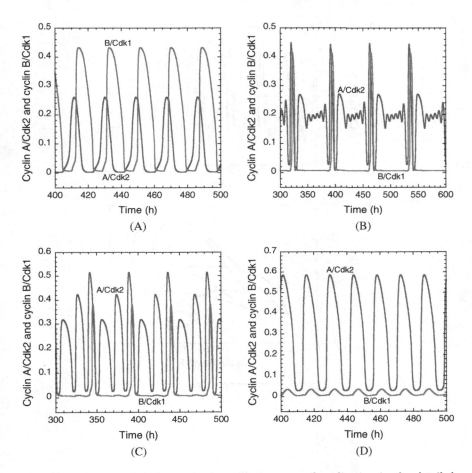

Fig. 5. Progressive shift from mitotic oscillations to endoreplication in the detailed model for the Cdk network.[33] From (A) to (D) the rate constant for inactivation of Cdh1 decreases progressively to yield (A) simple periodic, large-amplitude oscillations in cyclin A/Cdk2 and cyclin B/Cdk1); (B) complex periodic oscillations particularly noticeable for cyclin A/Cdk2; (C) yet another type of complex periodic oscillations; and (D) simple periodic oscillations corresponding to endoreplication, with large-amplitude oscillations in cyclin A/Cdk2 and small-amplitude oscillations in cyclin B/Cdk1 (see Ref. 33 for further details).

space these aperiodic oscillations correspond to the evolution to a strange attractor (Fig. 7C). Chaotic behavior is characterized by its sensitivity to initial conditions (Fig. 7B) and by the structure of the associated Poincaré section (Fig. 7D), which differs from that shown in Fig. 6C for complex periodic oscillations.

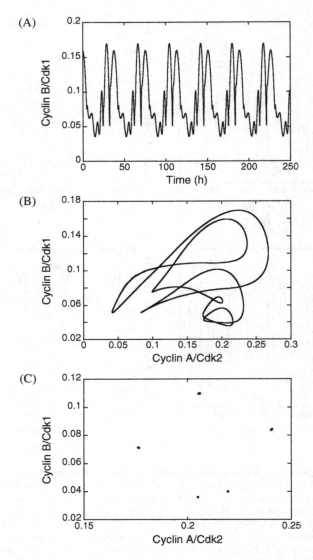

Fig. 6. Complex periodic oscillations in the detailed model for the Cdk network.[33] (A) Time evolution of cyclin B/Cdk1. (B) Projection of the trajectory of the 39-variable system onto the phase plane formed by cyclin A/Cdk2 and cyclin B/Cdk1. (C) Poincaré section established by plotting the level of cyclin B/Cdk1 versus the level of cyclin A/Cdk2 corresponding to the passage through a maximum in the level of cyclin E/Cdk2 (see Ref. 33 for further details).

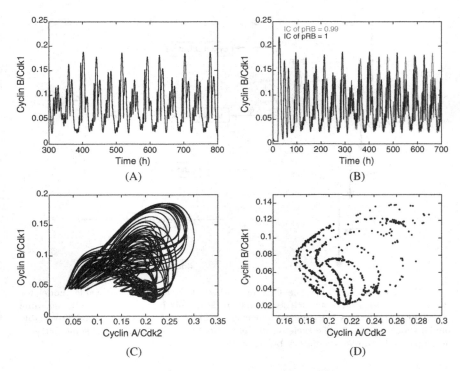

Fig. 7. Chaos in the detailed model for the Cdk network.[33] (A) Time evolution of cyclin B/Cdk1. (B) The time evolution of cyclin B/Cdk1 is shown again, for two slightly different initial values of pRB. The sensitivity to initial conditions is a signature of chaotic behavior. (C) Strange attractor obtained by projection of the trajectory of the 39-variable system onto the phase plane formed by cyclin A/Cdk2 and cyclin B/Cdk1. (D) Poincaré section established by plotting the level of cyclin B/Cdk1 versus the level of cyclin A/Cdk2 corresponding to the passage through a maximum in the level of cyclin E/Cdk2 (see Ref. 33 for further details).

The results obtained in the 39-variable model for the mammalian cell cycle (Fig. 2) were recovered in a reduced, skeleton version of this model, which contains only 5 variables.[34] This skeleton model, schematized in Fig. 8, disregards many biochemical details of the full model but retains its regulatory structure based on the four interacting cyclin/Cdk modules. Several oscillatory circuits can again be found in this simplified network. They synchronize when tightly coupled, to produce simple periodic behavior. However, as in the full model, complex patterns of oscillatory behavior occur when the strength of the coupling between the oscillatory circuits progressively diminishes.[34] Because the number of variables and parameters is significantly decreased, it is easier to find domains in

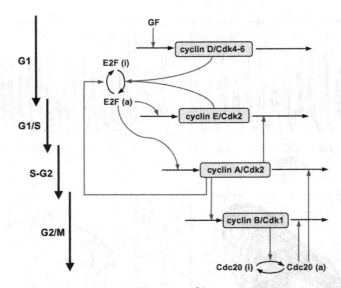

Fig. 8. Skeleton model for the Cdk network.[34] This reduced, 5-variable model for the Cdk network contains much less biochemical details but retains roughly the same regulatory interactions between the four cyclin/Cdk modules as the detailed 39-variable model shown in Fig. 2. The skeleton model is capable of reproducing the same patterns of simple and complex oscillatory behavior.[34]

parameter space where certain types of complex oscillations occur. Thus, birhythmicity was found numerically in the skeleton model, but it remains to be observed in the more detailed cell cycle model.

5. Comparison with Other Oscillatory Cellular Networks

Periodic behavior readily arises from feedback regulatory processes in biological systems. The nonlinearities associated with feedback regulation indeed give rise to instabilities and, consequently, favor the occurrence of oscillatory dynamics in cellular networks. The field of cellular rhythms has become a prototypic domain of research in systems biology, and such biochemical oscillatory phenomena have been the topic of successive reviews over the last decades.[1,19,23]

While oscillations generally take the form of simple periodic behavior corresponding to the evolution to a limit cycle in the phase space, more complex patterns of oscillatory behavior have been identified by means of modeling approaches: birhythmicity (i.e., the co-existence between two stable limit cycles separated by an unstable periodic trajectory), which is the periodic equivalent of bistability (i.e., the coexistence between two stable steady states); hard excitation (co-existence between a stable steady state and a stable limit cycle); complex periodic oscillations in the form of bursting; and aperiodic oscillations, i.e., chaos. Bursting oscillations are common in neurobiology, but are less frequently observed in non-electrically excitable cells, perhaps because in these cells they might not correspond to physiological conditions, which are more likely to produce simple periodic behavior. Here, using theoretical models, we focused on the conditions in which complex modes of oscillatory behavior of non-electrical nature arise in cellular networks.

To address the transition from simple to complex oscillatory behavior we started with a prototype model based on two autocatalytic enzyme reactions coupled in series. The interest of this simple, three-variable model is that it allows us to study the dynamical consequences of the interaction between two instability-generating mechanism, each of which is capable of producing oscillations. As recalled in Sec. 2, the interaction between the two oscillatory enzyme reactions markedly enhances the repertoire of oscillatory behavior. While a single autocatalytic enzyme reaction can only produce simple periodic oscillations, the system containing two coupled oscillatory reactions, in addition to limit cycle behavior, can display birythmicity, hard excitation, bursting, chaos, and even trirhythmicity.

The bienzymatic system was constructed as a prototype for investigating the interplay between two instability-generating mechanisms. But does such a situation arise in cellular networks and, if so, is it generic or relatively rare? The study of a large number of models for known cellular rhythms shows that the interplay between different endogenous oscillatory mechanisms is more the rule than the exception in cellular regulatory networks. This is illustrated by examples as diverse as those of cyclic AMP (cAMP) oscillations in *Dictyostelium* amoebae, intracellular Ca^{++} oscillations, circadian rhythms, the segmentation clock, and the cell cycle. The latter system was examined in detail in the two preceding sections. Before summarizing the results on the dynamics of the cell cycle, let us briefly examine the mechanism responsible for complex oscillations in networks that govern other cellular rhythms to see whether it also involves the interplay between two or more oscillatory circuits.

In *Dictyostelium* amoebae, oscillations of cAMP originate from the positive feedback exerted by extracellular cAMP on adenylate cyclase, the enzyme that synthesizes intracellular cAMP.[35] This self-amplification of cAMP synthesis follows the binding of extracellular cAMP to its membrane-bound receptor. The mechanism responsible for limiting the explosive production of cAMP relies on receptor desensitization: upon exposure to cAMP, the receptor undergoes a transition to an inactive (desensitized) state that is unable to elicit activation of adenylate cyclase. Such a mechanism can only produce simple periodic oscillations.[35] However, if one considers the variation of ATP, the substrate of adenylate cyclase, due to its consumption in the synthesis of cAMP, a second oscillatory mechanism is created: the two mechanisms involve the same positive feedback loop on adenylate cyclase but differ by the factor limiting self-amplification in cAMP synthesis; one mechanism relies on ATP consumption, while the other takes the form of receptor desensitization. The interaction of these two mechanisms acting in parallel to limit the effects of the same positive feedback loop is the source of bursting and chaos in the cAMP signaling system in *Dictyostelium*.[36,37]

The occurrence of complex Ca^{++} oscillations appears to originate from a different scenario. Simple periodic Ca^{++} spiking relies on a positive feedback regulation known as Ca^{++}-induced Ca^{++} release (CICR), itself triggered by the rise in inositol 1,4,5-trisphosphate (IP3) that follows stimulation of the cell by a hormone or neurotransmitter.[10,11,38] Rather than involving the interplay between two oscillatory mechanisms, the analysis of models indicates that oscillations of the bursting type and chaos result from the modulation by Ca^{++} oscillations of one of the parameters that control this oscillatory mechanism.[39,40] Thus, complex oscillations arise when IP3 becomes a variable and when its degradation is activated by Ca^{++}. Such periodic auto-modulation of a control parameter by the Ca^{++} oscillator itself can be viewed as the autonomous analogue of the non-autonomous forcing of an oscillatory system by the periodic variation of one of its parameters. Although the model for Ca^{++} signaling contains but a single oscillatory mechanism, the fact that its oscillatory output controls periodically the input parameter might be viewed as a first oscillator (the input) acting on and modulated by the second, core Ca^{++} oscillator.

That there are multiple ways to elicit the transition from simple to complex periodic oscillations is well illustrated by the case of Ca^{++} signaling. Besides the scenario mentioned above, two other mechanisms for producing complex Ca^{++} oscillations were identified in closely related models.[39] These extended versions, naturally, again involved at least three variables. In contrast, the original model for simple periodic Ca^{++} oscillations based on CICR contained only two variables.[38]

Birhythmicity, bursting and chaos were also observed in a realistic model for circadian rhythms in *Drosophila*.[41] The mechanism of the oscillations relies on a negative feedback loop exerted on their genes by a complex formed by the PER and TIM proteins. This negative feedback on gene transcription is responsible for the onset of sustained oscillations of the limit cycle type characterized by a period close to 24 h. What is the origin of complex oscillations, in view of the existence of a single oscillatory mechanism based on a unique negative feedback loop? The model reveals that two limbs contribute to the formation of the PER-TIM complex that exerts transcriptional inhibitory feedback (this negative feedback has since been shown to be mediated by the inhibition by PER-TIM of a complex of two activating proteins, CYCLE and CLOCK). Complex oscillations develop only when there is a slight imbalance between the processes of synthesis and degradation of the PER and TIM proteins.[41] This example provides another illustration that complex oscillations do not necessarily originate from the interplay between two oscillatory mechanisms or, as in the case of Ca^{++} oscillations, from periodic self-modulation of a parameter controlling the oscillations. However, the circadian situation might somehow be viewed as an interaction between two oscillators sharing, via the PER and TIM limbs, the same negative feedback loop, a situation reminiscent of the mechanism responsible for complex cAMP oscillations in *Dictyostelium*, as discussed above. Non-autonomous chaotic behavior can also be observed in a model for circadian rhythms driven by the light–dark cycle, as a result of the periodic variation of a light-sensitive parameter.[42]

A more detailed model for the mammalian circadian clock, containing up to 19 variables, was shown to harbor two distinct oscillatory mechanisms.[43,44] Nevertheless, numerical simulations so far brought evidence only for simple periodic behavior in this model. This by no means excludes the possibility of complex oscillatory behavior, if only because the parameter space was not explored in a systematic manner. The absence of bursting or chaos in the region explored numerically could mean either that one of the oscillatory circuits does not operate in the parameter domain corresponding to periodic behavior, or that the coupling between the two oscillatory circuits is so strong that they synchronize. Besides such internal synchronization, let us note that even if the circadian regulatory network operated in the domain of chaotic behavior, external synchronization with the light–dark cycle would likely suppress chaos and transform it into rhythmic behavior of 24-h period.

Discussed here in more detail, the model for the cyclin/Cdk network brings to light the various modes of dynamic behavior that emerge from the intertwined regulations of the different modules that drive the successive phases of the mammalian cell cycle. When the growth factor that induces progression in the cycle exceeds a critical value, repetitive activation of the cyclin/Cdk complexes occurs in the form of self-sustained oscillations. The periodic, ordered activation of the cyclin/Cdk complexes corresponds to the passage through the successive phases of the cell cycle. Much as in models containing but a few variables, such periodic oscillations correspond to the evolution to a limit cycle (Fig. 3) even though the system contains 39 variables.[15,27] Similar results are recovered in a skeleton, 5-variable version of the full model.[34] The Cdk network is thus capable of temporal self-organization in the form of limit cycle oscillations. This conclusion supports the view that the cell cycle operates on a limit cycle.[45]

Because of its relatively complex structure built along four Cdk modules and of the large number of variables and regulations involved, the Cdk network contains multiple oscillatory circuits. In spite of the multiplicity of oscillatory circuits, their tight coupling within the Cdk network ensures that periodic oscillations represent its most common mode of temporal self-organization. Periodic oscillations involving large-amplitude changes in Cdk2 and Cdk1 occur in large regions in parameter space, which often extend over several orders of magnitude of the control parameters.[15,27] It is when the coupling of the oscillatory circuits within the full Cdk network becomes weaker that their interactions may lead to complex patterns of oscillatory behavior. Distinct modes of complex oscillations can then be observed: bursting in the form of more or less complex periodic oscillations (Figs. 5 and 6), quasiperiodic oscillations, and chaos (Fig. 7). Poincaré sections established for these various modes of complex oscillatory behavior confer to each of them a distinct signature. Complex periodic oscillations occur in relatively smaller regions in parameter space as compared to simple periodic oscillations for which a single peak in Cdk1 and Cdk2 is observed per period. This observation holds with the view that the latter mode of periodic behavior represents the physiological mode of operation of the Cdk network, whereas complex oscillations, particularly quasiperiodic or chaotic behavior, would correspond to unphysiological or pathological conditions.

The model for the mammalian cell cycle allows us to investigate how the internal synchronization between multiple oscillatory circuits generally produces periodic behavior rather than chaos or quasiperiodic oscillations. The wiring of the regulatory network, rather than the number of variables or the degree of biochemical detail retained in the mathematical description, plays a key role in the onset of complex oscillations. This conclusion is supported by the observation of similar modes of complex oscillatory behavior in the full model for the mammalian cell cycle[15,27] and its reduced, skeleton version.[34] What is the general relevance of these results for the dynamics of cellular networks? The multiplicity of instability-generating mechanisms is widely encountered in cellular regulatory networks. Indeed, their complexity, due to the profusion of positive and negative feedback loops, is such that these networks often contain more than one source of oscillatory behavior. This is well illustrated by the circadian clock, the cell cycle, and the other examples of cellular rhythms mentioned above. Such multiplicity of interacting oscillatory circuits is also observed in the dynamics of yet another cellular rhythm, the segmentation clock, which is responsible for the periodic formation of somites in vertebrates.[13] In this remarkable example of spatio-temporal organization in embryogenesis, three oscillating pathways involving, respectively, Notch, Fgf, and Wnt signaling form the core of the oscillator,[46] even if the molecular mechanism of the segmentation clock remains to be fully characterized. In this system, numerical simulations indicate that complex periodic oscillations may originate from the interplay between three coupled signaling circuits,[47] each of which is capable of autonomous oscillations. However, in the case of the segmentation clock as in those of the cell cycle and circadian rhythms, and in contrast to bursting oscillations of the membrane potential in neurobiology, it appears that simple periodic oscillations remain the most frequent pattern of temporal organization and likely represent the physiological mode of oscillation in these regulatory networks, which control some of the most important cellular rhythms.

Acknowledgments

This work was supported by Grant No. 3.4607.99 from the Fonds de la Recherche Scientifique Médicale (F.R.S.M., Belgium).

References

1. A. Goldbeter, *Biochemical Oscillations and Cellular Rhythms: The Molecular Bases of Periodic and Chaotic Behaviour* (Cambridge Univ. Press, Cambridge, UK, 1996).
2. A. T. Winfree, *The Geometry of Biological Time* (Springer, New York, 2001).
3. A. Goldbeter, *La vie oscillatoire. Au cœur des rythmes du vivant* (Odile Jacob, Paris, 2010).
4. J. L. Feldman and C. A. Del Negro, *Nat. Rev. Neurosci.* **7**, 232–242 (2006).
5. G. Buzsaki, *Rhythms of the Brain* (Oxford University Press, New York, 2006).
6. D. Noble, *The Initiation of the Heartbeat* (Oxford Univ. Press, Oxford, 1979).
7. B. Hess and A. Boiteux, *Annu. Rev. Biochem.* **40**, 237–258 (1971).
8. A. Goldbeter and S. R. Caplan, *Annu. Rev. Biophys. Bioeng.* **5**, 449–476 (1976).
9. M. F. Madsen, S. Dano and P. G. Sorensen, *FEBS J.* **272**, 2648–2660 (2005).
10. M. J. Berridge, *Biochim. Biophys. Acta* **1793**, 933–940 (2009).
11. G. Dupont, L. Combettes, G. Bird and J. W. Putney, *Cold Spring Harb. Perspect. Biol.* **3**(3), a004226 (2010).
12. G. Gerisch and U. Wick, *Biochem. Biophys. Res. Comm.* **65**, 364–370 (1975).
13. O. Pourquié, *Science* **301**, 328–330 (2003).

14. D. O. Morgan, *The Cell Cycle: Principles of Control* (Oxford Univ. Press, Oxford, UK, 2006).
15. C. Gérard and A. Goldbeter, *Proc. Natl. Acad. Sci. USA* **106**, 21643–21648 (2009).
16. C. Dibner, U. Schibler and U. Albrecht, *Annu. Rev. Physiol.* **72**, 517–549 (2010).
17. C. L. Baker, J. J. Loros and J. C. Dunlap, *FEMS Microbiol. Rev.* **36**, 95–110 (2012).
18. M. Nakajima, K. Imai, H. Ito, T. Nishiwaki, Y. Murayama, H. Iwasaki, T. Oyama and T. Kondo, *Science* **308**, 414–415 (2005).
19. A. Goldbeter, C. Gérard, D. Gonze, J.-C. Leloup and G. Dupont, *FEBS Lett.* **586**, 2955–2965 (2012).
20. A. Goldbeter, *Adv. Chem. Phys.* **135**, 253–295 (2007).
21. W. B. Adams and J. A. Benson, *Progr. Biophys. Mol. Biol.* **46**, 1–49 (1985).
22. R. J. Field and L. Györgyi, *Chaos in Chemistry and Biochemistry* (World Scientific Publ., Singapore, 1993).
23. A. Goldbeter, *Nature* **420**, 238–245 (2002).
24. O. Decroly and A. Goldbeter, *Proc. Natl. Acad. Sci. USA* **79**, 6917–6921 (1982).
25. A. Goldbeter and R. Lefever, *Biophys. J.* **12**, 1302–1315 (1972).
26. O. Decroly and A. Goldbeter, *J. Theor. Biol.* **124**, 219–250 (1987).
27. C. Gérard and A. Goldbeter, *Front. Physiol.* **3**, 413 (2012). doi:10.3389/fphys.2012.00413.
28. N. Zielke, K. J. Kim, V. Tran, S. T. Shibutani, M. J. Bravo, S. Nagarajan, M. van Straaten, B. Woods, G. von Dassow, C. Rottig, C. F. Lehner, S. S. Grewal, R. J. Duronio and B. A. Edgar, *Nature* **480**, 123–127 (2011).
29. C. S. Sorensen, C. Lukas, E. R. Kramer, J.-M. Peters, J. Bartek and J. Lukas, *Mol. Cell. Biol.* **20**, 7613–7623 (2000).
30. B. Novak and J. J. Tyson, *Proc. Natl. Acad. Sci. USA* **94**, 9147–9152 (1997).
31. A. Csikasz-Nagy, D. Battogtokh, K. C. Chen, B. Novak and J. J. Tyson, *Mol. Biol. Cell* **19**, 3426–3441 (2006).
32. J. R. Pomerening, J. A. Ubersax and J. E. Ferrell Jr, *Mol. Biol. Cell* **19**, 3426–3441 (2008).
33. C. Gérard and A. Goldbeter, *Chaos* **20**, 045109 (2010).
34. C. Gérard and A. Goldbeter, *Interface Focus* **1**, 24–35 (2011).
35. J. L. Martiel and A. Goldbeter, *Biophys. J.* **52**, 807–828 (1987).
36. J. L. Martiel and A. Goldbeter, *Nature* **313**, 590–592 (1985).
37. J. L. Martiel and A. Goldbeter, *Lect. Notes Biomath.* **71**, 244–255 (1987).
38. A. Goldbeter, G. Dupont and M. J. Berridge, *Proc. Natl. Acad. Sci. USA* **87**, 1461–1465 (1990).
39. J. Borghans, G. Dupont and A. Goldbeter, *Biophys. Chem.* **66**, 25–41 (1997).
40. G. Houart, G. Dupont and A. Goldbeter, *Bull. Math. Biol.* **61**, 507–530 (1999).
41. J. C. Leloup and A. Goldbeter, *J. Theor. Biol.* **198**, 445–459 (1999).
42. D. Gonze and A. Goldbeter, *J. Stat. Phys.* **101**, 649–663 (2000).

43. J. C. Leloup and A. Goldbeter, *Proc. Natl. Acad. Sci. USA* **100**, 7051–7056 (2003).
44. J. C. Leloup and A. Goldbeter, *J. Theor. Biol.* **230**, 541–562 (2004).
45. C. Gérard and A. Goldbeter, *Math. Model. Nat. Phenom.* **7**, 126–166 (2012).
46. A. Aulehla and O. Pourquié, *Curr. Opin. Cell. Biol.* **20**, 632–637 (2008).
47. A. Goldbeter and O. Pourquié, *J. Theor. Biol.* **252**, 574–585 (2008).

Chapter 2

TIME-DEPENDENT MICHAELIS–MENTEN EQUATIONS FOR OPEN ENZYME NETWORKS

Jon Young*, Dieter Armbruster* and John Nagy*,[†],[‡]

*School of Mathematical and Statistical Sciences,
Arizona State University, Tempe, AZ, USA

[†]Department of Life Sciences,
Scottsdale Community College Scottsdale, AZ, USA

[‡]Physical Sciences Oncology Center,
Arizona State University, AZ, USA

1. Introduction

The derivation of the Michaelis–Menten equation and the experiments to determine the steady-state parameters for a basic enzyme-substrate reaction have been in standard undergraduate textbooks on chemistry for decades. In fact, most of the kinetic parameters listed in databases for enzymes are the steady-state, Michaelis–Menten parameters. However, most of the work validating the Michaelis–Menten approximation has been done on closed-systems, i.e., reactions with no influx or out-flux of molecules. This is not the appropriate model for open enzyme networks or for enzyme reactions that are embedded in larger networks which lead to oscillations or other dynamical behavior of interest. To study these issues we derive an open-Michaelis–Menten approximation using perturbation techniques similar to the ones used for the closed network. As an example we examine a basic enzymatic-cascade with time-dependent input where the product of one enzyme-substrate reaction becomes the enzyme in the next downstream enzyme-substrate reaction.

2. The Michaelis–Menten Equation

One of the most basic chemical reactions studied is the enzyme-substrate reaction. Early in the 20th century, the mechanism in which an enzyme converts a substrate into a product molecule was not fully understood. In 1913, Leonor Michaelis and Maud Menten proposed a mass-action model where the enzyme binds to the substrate to form an intermediate complex molecule.[1,2] The complex then either disassociates back into an enzyme and substrate, or the enzyme can successfully catalyze the conversion of substrate to product. The representative stoichiometry is

$$E + S \underset{d_1}{\overset{a_1}{\rightleftharpoons}} C \xrightarrow{k_1} E + P,$$

where a_1, d_1, and k_1 are rate constants. The corresponding system of equations is

$$\dot{E} = -a_1 ES + (d_1 + k_1)C, \tag{1a}$$

$$\dot{S} = -a_1 ES + d_1 C, \tag{1b}$$

$$\dot{C} = a_1 ES - (d_1 + k_1)C, \tag{1c}$$

$$\dot{P} = k_1 C, \tag{1d}$$

$$E(0) = E_0, \quad S(0) = S_0, \quad C(0) = P(0) = 0, \tag{1e}$$

where the initial conditions describe an initial concentration of enzymes and substrates where no reaction has yet occurred. Since no molecules are added or destroyed from the system, and also by examining (1), it is clear that certain quantities are conserved. Hence, (1) is equivalent to a system of two Ordinary Differential Equations (ODEs) and two algebraic equations.

$$E = E_0 - C, \tag{2a}$$

$$P = S_0 - (S + C), \tag{2b}$$

$$\dot{S} = -a_1(E_0 - C)S + d_1 C, \tag{2c}$$

$$\dot{C} = a_1(E_0 - C)S - (d_1 + k_1)C, \tag{2d}$$

$$S(0) = S_0, \quad C(0) = 0. \tag{2e}$$

Michaelis and Menten also proposed a method in which Eq. (2) can be approximated by a single ODE. In 1925, G. E. Briggs and J. B. S Haldane refined that approach based on the assumption that the complex does not change much during the timescale of the substrate depletion.[3] This can be achieved if the number of enzymes are much larger than the number of substrates, since any free enzyme will quickly bind to a free substrate. After a quick transient period in which the substrate concentration has not decayed much, $\dot{C} \approx 0$. This is known as the standard quasi-steady state assumption (sQSSA). If $\dot{C} = 0$, then C can be solved for in Eq. (2d),

$$C = \frac{E_0 S}{K_m + S}, \tag{3}$$

where $K_m = \frac{d_1 + k_1}{a_1}$ is the Michaelis–Menten constant. Then Eq. (3) can be substituted in Eq. (2c) to obtain

$$\dot{S} = \frac{-V_{\max} S}{K_m + S}, \quad S(0) = S_0. \tag{4}$$

Here $V_{\max} = k_1 E_0$ is the maximal possible rate at which the product could be formed and K_m can be regarded as the substrate concentration at which the formation rate of the product is half of $V_{\max}$. The initial condition in Eq. (4) assumes that the substrate has not decayed noticeably during the initial transient period. Equation (4) is generally referred to as the Michaelis–Menten equation. In Refs. 4 and 5 it was shown that Eq. (4) has a closed form solution

$$S(t) = K_m W \left[\frac{S_0}{K_m} \exp \left(\frac{S_0}{K_m} - \frac{V_{\max}}{K_m} t \right) \right],$$

where W is the Lambert W function. Experimentally, it easier to obtain the parameters K_m and $V_{\max}$ than it is to measure the reaction rate constants. Hence one typically finds the steady-state Michaelis–Menten parameters in the literature rather than the dynamic parameters of the mass action equations.

3. Perturbation Analysis of the Michaelis–Menten Equations

Since Eq. (4) is an approximation, one would like to have a mathematically precise way to evaluate the accuracy of this approximation and the errors made when using the approximation, as a function of the system parameters. The standard approach to answer such questions is to derive the approximations as a low order expansion in a perturbation theory. References 6 and 7 are excellent texts describing perturbation methods and applications. Specifically, in the case of the Michaelis–Menten equation, singular perturbation theory has been used, starting with scaling Eq. (2) and then identifying a small parameter. Early approaches scaled C by E_0 and S by S_0 and defined a parameter $\epsilon = E_0/S_0$.[7,8] In 1989, Segel and Slemrod gave a more rigorous derivation and scaling of the system[9] introducing the following scaled variables for Eq. (2),

$$s = \frac{S}{S_0},$$

$$c = \frac{C}{\bar{C}},$$

$$\tau = \frac{t}{t_C},$$

$$T = \frac{t}{t_S},$$

where $\bar{C}$ is an estimate of the maximal complex concentration, t_C is an estimate for the fast timescale, and t_S is an estimate for the slow timescale and τ and T measure time on the fast and slow timescales, respectively. It is assumed that in the initial transient period, the substrate depletion is minimal and $S \approx S_0$. By substituting S_0 for S in (2d), it is easy to show that

$$C(t) = \bar{C}[1 - \exp(-kt)], \tag{5}$$

where

$$\bar{C} = \frac{E_0 S_0}{K_m + S_0}, \tag{6}$$

$$k = a_1(S_0 + K_m).$$

Using (5) to get an estimate for the fast time scale, one finds $t_C = k^{-1}$. The characterization of a time scale from Ref. 7 is used to estimate the slow timescale as

$$t_S = (S_{max} - S_{min}) \bigg/ \left|\frac{dS}{dt}\right|_{max}. \tag{7}$$

Estimating t_s by plugging S_0 into (4) one obtains:

$$t_S = \frac{K_m + S_0}{k_1 E_0}. \tag{8}$$

As a result of these scalings, the system depends on three dimensionless parameters:

$$\sigma = \frac{S_0}{K_m}, \quad \kappa = \frac{d_1}{k_1}, \quad \epsilon = \frac{E_0}{K_m + S_0}.$$

In our analysis we assume $\epsilon \ll 1$ and use it as the small parameter for a perturbation expansion, whereas σ and κ are assumed to be of $O(1)$. Then, on the fast timescale, the governing equations for (2) are:

$$s'(\tau) = \epsilon \left[-s + \frac{\sigma}{\sigma + 1} cs + \frac{\kappa}{(\kappa + 1)(\sigma + 1)} c \right],$$

$$c'(\tau) = s - \frac{\sigma}{\sigma + 1} cs - \frac{1}{\sigma + 1} c, \tag{9}$$

$$s(0) = 1, \quad c(0) = 0.$$

Expanding $s(\tau), c(\tau)$ in a powers of ϵ

$$s(\tau) \sim s^{(0)}(\tau) + \epsilon s^{(1)}(\tau) + \cdots, \quad c(\tau) \sim c^{(0)}(\tau) + \epsilon c^{(1)}(\tau) + \cdots,$$

we expand (9) into differential equations for each order of ϵ that can then be successively solved.

The solutions to the $O(1)$ problem on the fast timescale are:

$$s^{(0)}(\tau) = 1, \quad c^{(0)}(\tau) = 1 - e^{-\tau}. \tag{10}$$

On the slow timescale, the governing equations are:

$$s'(T) = (\kappa + 1)(\sigma + 1) \left[-s + \frac{\sigma}{\sigma + 1} cs + \frac{\kappa}{(\kappa + 1)(\sigma + 1)} c \right], \tag{11}$$

$$\epsilon c'(T) = (\kappa + 1)(\sigma + 1) \left[s - \frac{\sigma}{\sigma + 1} cs - \frac{1}{\sigma + 1} c \right].$$

Again, expanding $s(T), c(T)$ in a powers of ϵ, $s(T) \sim s_0(T) + \epsilon s_1(T) + \cdots$, $c(T) \sim c_0(T) + \epsilon c_1(T) + \cdots$, the governing equations for (11) become:

$$s_0'(T) + \epsilon s_1'(T) + \cdots$$

$$= (\kappa + 1)(\sigma + 1) \left[-s_0 + \frac{\sigma}{\sigma + 1} c_0 s_0 + \frac{\kappa}{(\kappa + 1)(\sigma + 1)} c_0 \right]$$

$$+ \epsilon (\kappa + 1)(\sigma + 1) \left[-s_1 + \frac{\sigma}{\sigma + 1} (c_0 s_1 + c_1 s_0) + \frac{\kappa}{(\kappa + 1)(\sigma + 1)} c_1 \right]$$

$$+ O(\epsilon^2),$$

$$0 + \epsilon c_0'(T) + \cdots$$

$$= (\kappa + 1)(\sigma + 1) \left[s_0 - \frac{\sigma}{\sigma + 1} c_0 s_0 - \frac{1}{\sigma + 1} c_0 \right]$$

$$+ \epsilon (\kappa + 1)(\sigma + 1) \left[s_1 - \frac{\sigma}{\sigma + 1} (c_0 s_1 + c_1 s_0) - \frac{1}{\sigma + 1} c_1 \right]$$

$$+ O(\epsilon^2).$$

The $O(1)$ equations become

$$c_0(T) = \frac{(\sigma + 1)s_0}{\sigma s_0 + 1}, \quad s_0'(T) = -c_0, \tag{12}$$

which are identical to the Michaelis–Menten equations when scaled back to the original variables. The $O(\epsilon)$ equations are found to be:

$$c_1(T) = \frac{s_1(1 + \sigma)}{(1 + \sigma s_0)^2} + \frac{(\sigma + 1)^2 s_0}{(1 + \kappa)(\sigma s_0 + 1)^4}, \tag{13a}$$

$$s_1'(T) = -c_1 - c_0' = -p(T)s_1 + q(T), \tag{13b}$$

where

$$p(T) = \frac{1 + \sigma}{(1 + \sigma s_0)^2}, \quad q(T) = \frac{(\sigma + 1)^2 s_0}{(\sigma s_0 + 1)^3}\left(1 - \frac{1}{(1 + \kappa)(\sigma s_0 + 1)}\right). \tag{14}$$

Segel and Slemrod use the $O(\epsilon)$ equations to determine the range of validity of their model of the Michaelis–Menten equation based on the $O(1)$ approximation. They find that, if the magnitudes of s_1 and c_1 are much larger than one, then perturbation analysis loses validity and the errors could be large when using the $O(1)$ approximation.

In 1996, Borghans[10] made the change of variables $S_c = S + C$ to extend the sQSSA model and derive the equivalent differential model for (2)

$$\dot{S}_c = -k_1 C, \tag{15a}$$

$$\dot{C} = a_1(E_0 - C)(S_c - C) - (d_1 + k_1)C, \tag{15b}$$

$$S_c(0) = S_0, \quad C(0) = 0. \tag{15c}$$

The *total quasi-steady state assumption, tQSSA*, states that Eq. (15b) is approximately equal to zero on the timescale of the total substrate depletion deriving a quadratic equation for C:

$$C^2 - (E_0 + K_m + S_c)C + E_0 S_c = 0. \tag{16}$$

Using the first Padé approximant to (16), $C_p = \frac{E_0 S_c}{E_0 + K_m + S_c}$ and substituting C_p for C into (15a) one obtains

$$\dot{S}_c = \frac{-V_{\max} S_c}{E_0 + K_m + S_c}. \tag{17}$$

By using a similar argument as used in Ref. 9, Borghans derived a different slow timescale

$$t_{sc} = \frac{E_0 + S_0 + K_m}{k_1 E_0}, \tag{18}$$

leading to a new perturbation parameter

$$\epsilon = \frac{k_1 E_0}{a_1(E_0 + S_0 + K_m)} \ll 1. \tag{19}$$

Observing Eq. (19) we find sufficient conditions for $\epsilon \ll 1$ to be

$$\frac{k_1}{a_1} \ll E_0 + S_0,$$

$$k_1 \ll d_1,$$

$$E_0 \ll S_0 + K_m.$$

Hence, if the sQSSA is valid, then so too is the tQSSA. However, the tQSSA greatly extends the regime of validity.

In 2007, Gomez-Uribe *et al.* used the tQSSA to study the operating regimes of a cyclic motif of covalent modification, such as seen in the phosphorylation and de-phosphorylation cycle in a MAPK cascade[11]:

$$E_1 + S \underset{d_1}{\overset{a_1}{\rightleftharpoons}} C_1 \xrightarrow{k_1} E_1 + P, \tag{20}$$

$$E_2 + P \underset{d_2}{\overset{a_2}{\rightleftharpoons}} C_2 \xrightarrow{k_1} E_2 + S.$$

By applying the tQSSA, the authors derived the approximation

$$\dot{P}_c = k_1 \frac{\bar{E}_1(\bar{S} - P_c(t))}{K_{m1} + \bar{E}_1 + \bar{S} - P_c(t)} - k_2 \frac{\bar{E}_2 P_c(t)}{K_{m2} + \bar{E}_2 + P_c(t)}, \tag{21}$$

where

$$P + C_2 = P_c, \quad S + C_1 + P + C_2 = \bar{S}, \quad E_1 + C_1 = \bar{E}_1, \quad E_2 + C_2 = \bar{E}_2,$$

and

$$K_{m1} = \frac{k_1 + d_1}{a_1}, \quad K_{m2} = \frac{k_2 + d_2}{a_2}.$$

Equation (21) have four different steady-state regimes. Additionally, by allowing $\bar{E}_1$ to be dependent on time one can study the dynamic response of (20). Studying the filtering characteristics of these equations by choosing $\bar{E}_1(t) = E_0(1 + a\sin(\omega t))$, Gomez-Uribe *et al.* discovered that the covalent modification cycle acts as a low-pass filter.

4. Basic Open-Enzyme Network

4.1. *Perturbation analysis*

The basic model of an open enzyme network has the stoichiometry

$$\varnothing \xrightarrow{\bar{\lambda}(t)} E,$$

$$E + S \underset{d_1}{\overset{a_1}{\rightleftarrows}} C \xrightarrow{k_1} E + P,$$

leading to the corresponding set of mass-action equations

$$\dot{E} = \bar{\lambda}(t) - a_1 ES + (d_1 + k_1)C,$$
$$\dot{S} = -a_1 ES + d_1 C,$$
$$\dot{C} = a_1 ES - (d_1 + k_1)C, \tag{22}$$
$$\dot{P} = k_1 C,$$
$$E(0) = C(0) = P(0) = 0, \quad S(0) = \bar{S}.$$

Mass-conservation and Borghans' coordinate transformation reduces the system to

$$S_C = S + C,$$
$$E + C = \int_0^t \bar{\lambda}(x)dx =: \bar{\Lambda}(t),$$
$$P = \bar{S} - S_C, \tag{23}$$
$$\dot{S}_C = -k_1 C,$$
$$\dot{C} = a_1(\bar{\Lambda} - C)(S_C - C) - (d_1 + k_1)C,$$
$$C(0) = 0, \quad S_C(0) = \bar{S}.$$

This is the fundamental system that we will analyze under time-dependent influx $\bar{\Lambda}(t)$. Note that $\bar{\Lambda}(t)$ represents the total enzyme concentration in the system, and hence can be assumed to be non-negative and bounded, i.e.,

$$\sup_{t \in [0,\infty)} \{\bar{\Lambda}(t)\} = \bar{E} < \infty.$$

Following the approach of Segel and Slemrod, we will scale the total substrate concentration S_C and the product P, identify a small parameter ϵ and apply singular perturbation theory.

To get an estimate for the maximum of C to provide a scaling parameter, it will be necessary to first show that such a maximum exists.

This can be argued physically, or by examining Eq. (23). The proof can be found in Ref. 12. Let $t_{\max}$ be the time at which the maximum occurs. Then

$$\dot{C}(t_{\max}) = 0 = a_1(\bar{\Lambda}(t_{\max}) - C(t_{\max}))S(t_{\max}) - (d_1 + k_1)C(t_{\max}).$$

This implies

$$C(t_{\max}) = \frac{\bar{\Lambda}(t_{\max})S(t_{\max})}{K_m + S(t_{\max})} \leq \frac{\bar{E}\bar{S}}{K_m + \bar{S}} =: \bar{C}, \tag{24}$$

where

$$K_m = \frac{d_1 + k_1}{a_1} \tag{25}$$

is the Michaelis–Menten parameter.

The timescale on which the total-substrate concentration operates on can be estimated by:

$$\frac{(S_{C_{\max}} - S_{C_{\min}})}{\left|\dot{S}_C\right|_{\max}} \approx \frac{\bar{S}}{k_1\bar{C}} = \frac{K_m + \bar{S}}{k_1\bar{E}} := t_{S_C}. \tag{26}$$

The variables in system (23) can thus be scaled as:

$$T = \frac{t}{t_{S_C}}, \quad s_c(T) = \frac{S_C(t)}{\bar{S}}, \quad c(T) = \frac{C(t)}{\bar{C}}, \quad p(T) = \frac{P(t)}{\bar{S}}, \quad \Lambda(T) = \frac{\bar{\Lambda}(t)}{\bar{E}},$$

leading again to three dimensionless parameters

$$\sigma = \frac{\bar{S}}{K_m}, \quad \kappa = \frac{d_1}{k_1}, \quad \epsilon = \frac{\bar{E}}{K_m + \bar{S}}.$$

Hence, after scaling, the system (23) is equivalent to the dimensionless system

$$\begin{aligned}
\Lambda(T) &= \frac{1}{\bar{\bar{E}}} \int_0^{t_s T} \bar{\lambda}(x)dx, \\
p(T) &= 1 - s(T), \\
s_c'(T) &= -c, \\
\epsilon c'(T) &= (\kappa + 1)\left[((\sigma + 1)\Lambda(T) - \sigma c)(s_c - \epsilon c) - c\right], \\
s_c(0) &= 1, \quad c(0) = 0.
\end{aligned}$$

$$\tag{27}$$

Assuming $\epsilon \ll 1$ and expanding s_c and c in powers of ϵ, $s_c(T) \sim s_{c_0}(T) + \epsilon s_{c_1}(T) + \cdots$, $c(T) \sim c_0(T) + \epsilon c_1(T) + \cdots$, the $O(1)$ equations become

$$s'_{c_0} = -c_0,$$

$$c_0 = \frac{(\sigma+1)\Lambda(T) s_{c_0}}{\sigma s_{c_0} + 1}, \tag{28}$$

$$s_{c_0}(0) = 1, \quad c_0(0) = 0.$$

The system (28) has an explicit solution:

$$s_{c_0}(T) = \frac{1}{\sigma} W\left[\sigma \exp\left(\sigma - (1+\sigma)\int_0^T \Lambda(x)dx\right)\right],$$

$$p_0(T) = 1 - \frac{1}{\sigma} W\left[\sigma \exp\left(\sigma - (1+\sigma)\int_0^T \Lambda(x)dx\right)\right], \tag{29}$$

where W is the Lambert W function.

4.2. *Time-dependent Michaelis–Menten equations for a larger network*

We apply the approach outlined in the last section to study an enzymatic cascade. By examining the properties of the Lambert W function and the exponential function, it can be shown that the functional operator:

$$F : (\Lambda_{\text{set}}, \mathbb{R}_{>0}) \to \Lambda_{\text{set}},$$

defined as:

$$F_\sigma(\Lambda) = 1 - \frac{1}{\sigma} W\left[\sigma \exp\left(\sigma - (1+\sigma)\int_0^T \Lambda(x)dx\right)\right] \tag{30}$$

is well-defined for Λ_{set} the appropriate set of functions of non-negative and bounded influxes. Let

$$F^1(\Lambda) = F_{\sigma_1}(\Lambda), \quad F^2(\Lambda) = F_{\sigma_2}(F_{\sigma_1}(\Lambda)), \quad \cdots.$$

Then the output for an n-stage cascade at the $O(1)$ perturbation level is approximated by:

$$p_0^n(T) = F^n(\Lambda)(T).$$

In Ref. 13 we study this n-stage cascade as a generic model for signal transduction cascades. We show that iterating the operator defined in Eq. (30) three times creates a reliable sigmoidal response progress curve from a wide variety of time-dependent signaling inputs suggesting that natural selection may have favored signaling cascades as a parsimonious solution to the problem of generating switch-like behavior in a noisy environment.

4.3. *Accuracy of the perturbation scheme*

We follow the approach introduced by Segel and Slemrod and study the $O(\epsilon)$ expansion of the system (27) to the first order approximation of the time-dependent Michaelis–Menten system (28). The $O(\epsilon)$ equations are:

$$c_1 = \frac{1}{1 + \sigma s_{c_0}} \left(c_0(\sigma c_0 - (\sigma + 1)\Lambda) + ((\sigma + 1)\Lambda - \sigma c_0)s_{c_1} - \frac{c_0'}{\kappa + 1} \right),$$

$$s_{c_1}' = -c_1,$$

$$s_{c_1}(0) = 0.$$

Since s_{c_0} and c_0 have been solved explicitly at $O(1)$ they are given functions of T and hence the equation for s_{c_1} is a first order time-dependent linear differential equation of the form

$$s_{c_1}' = -P(T)s_{c_1} + Q(T),$$

that can be solved by the method of integrating factors. After some algebraic manipulations (details see Ref. 12) we show that the integrating factor can be written as

$$\mu(T) = \frac{1 + \sigma s_{c_0}}{(\sigma + 1)s_{c_0}}.$$

Hence,

$$\int_0^T \mu(y)Q(y)dy = \int_0^T \frac{\Lambda'}{(\kappa + 1)(\sigma s_{c_0} + 1)}$$

$$+ \Lambda \left(\frac{s_{c_0}'}{(\kappa + 1)s_{c_0}(\sigma s_{c_0} + 1)^2} - \frac{s_{c_0}'}{s_{c_0}(\sigma s_{c_0} + 1)} \right) dy$$

and since $s_{c_0}(y)$ and $\Lambda(y)$ are scaled to lie between 0 and 1, we can bound the $O(\epsilon)$ term of the substrate by the expression

$$|s_{c_1}| \leq \frac{(\sigma + 1)s_{c_0}\Lambda}{(\sigma s_{c_0} + 1)(\kappa + 1)} + \frac{\sigma s_{c_0}(1 - s_{c_0})}{(\kappa + 1)(\sigma s_{c_0} + 1)^2}$$

$$- \frac{(\sigma + 1)s_{c_0}}{(\kappa + 1)(\sigma s_{c_0} + 1)} \log \left(\frac{(\sigma + 1)s_{c_0}}{\sigma s_{c_0} + 1} \right)$$

$$-\frac{(\sigma+1)s_{c_0}}{\sigma s_{c_0}+1}\log\left(\frac{(\sigma+1)s_{c_0}}{\sigma s_{c_0}+1}\right)$$

$$\leq \frac{1}{\kappa+1}\left(2+\frac{1}{e}\right)+\frac{1}{e}<3.$$

Therefore if $\epsilon \ll 1$, then $\epsilon|s_{c_1}| \ll 1$ and the perturbation scheme is valid. Note however, that we do not have a similar bound for the error of the intermediate complex c which depends on Λ' and σ. In particular, if Λ' is large, then the error in c could be large.

Since F is an $O(1)$ approximation to the true output, treating a cascade as an n-fold iteration of F has the potential to introduce additional error. It is intuitive that one can trade the number of iterations against the smallness of ϵ.

We can see how perturbations in the input would propagate through the approximated model. Suppose Λ is the input and $\Lambda + \epsilon\Lambda_1$ is the perturbed input. Let s_{c_0} be the output for Λ and $\tilde{s}_{c_0}$ be the output to the perturbed input. Then

$$\log(s_{c_0})+\sigma s_{c_0}=\sigma-(\sigma+1)\int_0^T \Lambda(x)dx,$$

$$\log(\tilde{s}_{c_0})+\sigma\tilde{s}_{c_0}=\sigma-(\sigma+1)\int_0^T \Lambda(x)+\epsilon\Lambda_1(x)dx,$$

which implies

$$\log(s_{c_0})-\log(\tilde{s}_{c_0})+\sigma(s_{c_0}-\tilde{s}_{c_0})=(\sigma+1)\epsilon\int_0^T \Lambda_1(x)dx.$$

By the Mean Value Theorem, there exists $\xi \in (\tilde{s}_{c_0}, s_{c_0}) \subset (0,1]$, such that

$$\log(s_{c_0})-\log(\tilde{s}_{c_0})=\frac{1}{\xi}(s_{c_0}-\tilde{s}_{c_0}).$$

This implies that

$$|s_{c_0}-\tilde{s}_{c_0}|=\left|\frac{\sigma+1}{\sigma+1/\xi}\right|\left|\epsilon\int_0^T \Lambda_1(x)dx\right|\leq\epsilon\left|\int_0^T \Lambda_1(x)dx\right|.$$

This suggests that using a sequence of function compositions as a model of a signaling cascade works well if $\left|\int_0^\infty s_1(x)dx\right|$ is bounded. This is in particular true for an influx $\lambda(t)$ that goes to zero at a time of $O(1)$ on the slow timescale — a reasonable biological assumption.

4.4. *Simulation results and discussion*

For the biologically relevant situations that we checked through simulations the outputs tend to stay close to each other. Figure 1 demonstrates the validity of the perturbation scheme even with a periodic input. Figure 2 shows that errors do tend to accumulate, but the rate appears linearly dependent on ϵ. However, we lack a complete mathematical proof of these two facts. In particular, restricting the input $\Lambda(T)$ to the natural set of nonnegative and bounded continuous functions is not enough: Using the fact that $\left| \int_0^\infty s_1(x)dx \right|$ can be made relatively large one can contrive counterexamples demonstrating a large error between the outputs with inputs that

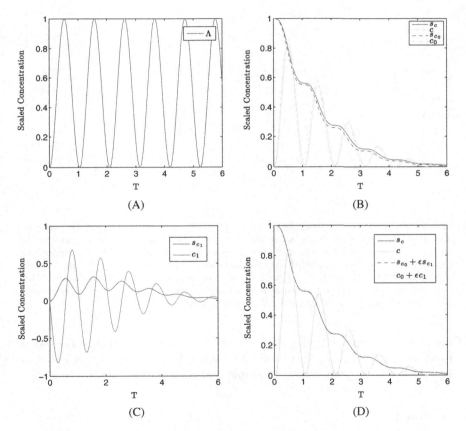

Fig. 1. Plots demonstrating the accuracy of the perturbation expansion. In (a), the scaled input $\Lambda = (1 - \cos(\omega T))/2$ with $\omega = 6$ is plotted. Equation (27) and its $O(1)$ and $O(\epsilon)$ approximations were integrated with $\sigma = 1$, $\kappa = 1$, and $\epsilon = 0.1$.

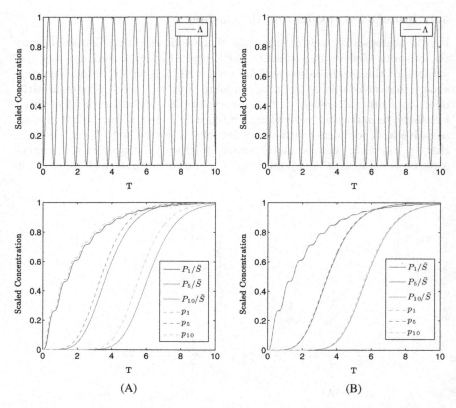

Fig. 2. Plots demonstrating the accuracy of approximating an enzymatic cascade. The first row shows that the same input function is used; $\bar{\Lambda} = \bar{E}(1 - \cos(\omega t))/2$ with $\omega = 10\,\text{min}^{-1}$. Equation (27) was integrated and scaled and then compared to iterating Eq. (30). For simplicity, the rate parameters were derived from the parameters $\bar{E} = 0.5\,\mu M$, $\bar{S}_i = 0.5\,\mu M$, $\kappa = 1$, and $t_{s_c} = 1$. In column (a), $\epsilon = 0.1$. In column (b), $\epsilon = 0.01$. It appears that the errors tend to accumulate, but more work is needed to determine exactly at what rate the accumulation occurs.

are $O(\epsilon)$ between each other. More work is needed to determine the exact function space for $\Lambda(T)$ that would guarantee close outputs in cascades and bounds for error propagation in simple and complex enzymatic systems.

5. Conclusion

Much work has been done over the years in analyzing mass-action models of enzymatic reactions. Numerous studies have been made that examined the Michaelis–Menten approximation of closed enzymatic systems, and some studies have been made towards analyzing perturbation approximations of open enzymatic reactions; however, this is the first work to rigorously show that Michaelis–Menten parameters derived from closed experiments are applicable to open enzymatic systems under certain conditions. It was argued in Ref. 14 that many researchers use the Michaelis–Menten approximation without first validating the conditions under which the approximation is valid. One such situation that has not been looked at in detail is the use of the Michaelis–Menten approximation for enzyme-substrate reactions when those reactions are embedded in larger chemical networks. Our result supports the practice to use the Michaelis–Menten approximation in time-dependent systems which has been done in networked modules without proper justification. This will be a major boon to systems biologists since most parameters for enzymatic reactions are derived from isolated experiments.

This work validated the Michaelis–Menten approximation when $\epsilon = \bar{E}/(\bar{S} + K_m) \ll 1$. More work can be done to see what happens when $\epsilon = O(1)$ or when $\epsilon \gg 1$. Models with feedback, an influx of substrate, and a combination of other tweaks, will also be examined in the future.

Beyond the validity of the Michealis–Menten approximation for a one stage substrate-enzyme reaction our research also highlighted the issue of error accumulation in models of complex enzyme reaction models where a characterization of one reaction via the Michaelis–Menten approximation leads to an error for the influx of enzymes in a neighboring node of the network. If the trigger signal of the enzymatic network is fast and travels through the network fast, then we showed that the error stays bounded at least for an enzymatic cascade. However, for a periodic input on an infinite time interval, errors will accumulate. More work needs to be done to examine how errors will propagate and accumulate for long signals or for signals that travel slowly through a complicated network of chemical reactions.

Acknowledgments

This work was partially supported by More Graduate Education at Mountain States Alliance (MGE@MSA), Alliance for Graduate Education and the Professoriate (AGEP), National Science Foundation (NSF) Cooperative Agreement No. HRD-0450137 (to J.Y.) and by the Volkswagen Foundation under the program on Complex Networks (to D.A. and J.Y.).

References

1. L. Michaelis and M. L. Menten, *Biochem. Z.* **49**, 333–369 (1913).
2. K. A. Johnson and R. S. Goody, *Biochemistry* **50**(39), 8264–8269 (2011).

3. G. E. Briggs and J. Haldane, *Biochem. J.*, **19**(2), 338–339 (1925).
4. S. Schnell and C. Mendoza, *J. Theor. Biol.* **187**, 207–212 (1997).
5. R. M. Corless *et al.*, *Adv. Comput. Math.* **5**(1), 329–359 (1996).
6. M. H. Holmes, *Introduction to Perturbation Methods* (Springer Verlag New York, Inc., 1995).
7. C. C. Lin and L. A. Segel, *Mathematics Applied to Deterministic Problems in Natural Sciences* (Macmillan Publishing, 1974).
8. F. Heineken, H. Tsuchiya and R. Aris, *Math. Biosci.* **1**(1), 95–113 (1967).
9. L. A. Segel and M. Slemrod, *SIAM Rev.* **31**(3), 446–447 (1989).
10. J. Borghans *et al.*, *Bull. Math. Biol.* **58**, 43–63 (1996).
11. C. Gomez-Uribe *et al.*, *PLoS Comput. Biol.* **3**(12), 2487–2497 (2007).
12. J. Young, *Time-Dependent Models of Signal Transduction Networks*, Ph.D. thesis, Arizona State University (2013).
13. J. Young, D. Armbruster and J. Nagy, Three level signal transduction cascades lead to reliable switches, preprint, Arizona State University (2014).
14. A. M. Bersani and G. Dell'Acqua, *J. Math. Chem.* **50**, 335–344 (2012).

Chapter 3

ENVIRONMENTAL DEPENDENCE
OF THE ACTIVITY AND ESSENTIALITY
OF REACTIONS IN THE METABOLISM
OF *ESCHERICHIA COLI*

Oriol Güell[*,‡], M. Ángeles Serrano[†] and Francesc Sagués[*]

*Departament de Química Física, Universitat de Barcelona,
Martí i Franquès 1, 08028 Barcelona, Spain*

†*Departament de Física Fonamental, Universitat de Barcelona,
Martí i Franquès 1, 08028 Barcelona, Spain*

‡*oguell@ub.edu*

1. Introduction

Metabolic networks are complex systems which contain hundreds of metabolites that interact through a large number of reactions.[1-3] Internal or external perturbations may modify their behavior.[4-7] In general, it has been found that complex networks are robust against accidental failures, but fragile against targeted attacks to their most connected important nodes.[8] In the case of metabolic networks, robustness has been studied in relation to failure of reactions.[9-11] In particular, resistance to lethality in metabolic networks can be related with the capability of a metabolic network to secure a nonzero growth rate. Hence, when the failure of a reaction leads to no growth conditions in a organism, this reaction is considered essential.

In cell biology, the concept of essentiality is generally referred to genes. However, it can be extended to reactions due to the fact that the knockout of a gene is translated into the inability of the metabolic network to use the reaction which is under control of the knocked out gene. This allows to compute cores of reactions which are important for metabolism,[12] or to extend the concept of essentiality into superessentiality, a property which will be explained in the following section.[13]

1.1. *Structure: Metabolism as a complex network*

A metabolic network is a complete set of chemical reactions which secures the life of an organism. Metabolism is responsible for the growth of organisms, maintaining structures, constructing new ones and responding to the surroundings.[14] Metabolism is divided in two basic categories, depending on the way it works. Catabolism is the term used when metabolism is in charge of breaking large molecules into small ones, like the oxidation of fatty acids. On the contrary, the used term when metabolism act as a builder of large molecules from small ones is anabolism. Examples of anabolism are the synthesis of proteins, lipids or amino acids. Metabolic reactions are catalyzed by proteins called enzymes, which are macromolecules composed of amino acids. Without enzymes, biochemical reactions would not take place, despite being thermodynamically spontaneous, since enzymes decrease the necessary activation energy of reactions and in this way rates of reactions increase. At the same time, enzymes are regulated by genes, which are DNA fragments with the necessary hereditary information. This gene-enzyme-reaction coupling is shown in Fig. 1.

Metabolic networks have been studied in the context of complex networks.[1,4,15] Their structures are represented as graphs, where nodes are metabolites and reactions. This implies that two kinds of nodes are needed

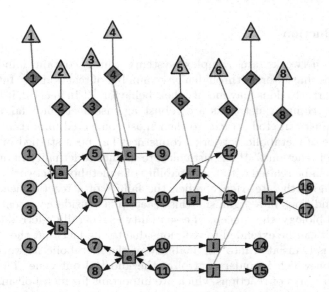

Fig. 1. Example of a directed metabolic network with enzymes catalyzing reactions and genes regulating enzymes. Reactions are represented by squares, metabolites by circles, enzymes by rhombus and genes by triangles.

for a complete representation of a metabolic network, leading to a bipartite representation where links only connect nodes within different categories.

Information about the reactions permits us to construct a directed representation of the network, where one can make a distinction between reactants and products and between reversible and irreversible reactions. With this information, the obtained network is more accurate and describes the system in a more realistic way.

Nevertheless, simpler representations of metabolic networks can be adopted in terms of unipartite networks, also called one-mode projections. In that version, networks have just one type of nodes. To construct them, typically metabolites are chosen as nodes, and a link is placed between two metabolites if there is at least one reaction which relates them. On the contrary, if reactions are chosen as nodes, two reactions are connected if there is at least a common metabolite between them.

As graphs, metabolic networks show the following characteristic features:

- Nodes are characterized by the number of neighbors. This magnitude is called the *degree*, k, of a node. Note that for directed networks, the degrees are divided into *in* degree, *out* degree and *bidirectional* degree, with their associated degree distributions.[16] The distribution of degrees $P(k)$ gives the probability that a node selected at random has a degree k.

   — In most cases, metabolites show a power-law degree distribution

$$P(k_M) \propto k_M^{-\beta} \tag{1}$$

   with a characteristic exponent β. For most complex networks in the real world, β has values in the range $2 < \beta < 3$. For metabolic networks, β is typically ~ 2.4. The fact that metabolites display a scale-free distribution means that there is a high diversity in the degrees of the nodes of the network. Networks with a degree distribution described by a power-law are called scale-free. As a consequence, most metabolites have few connections whereas a few metabolites, called hubs, have many of them. An example of these highly-connected metabolites is ATP, an important currency metabolite which is involved in many reactions as an energy contributor.

   — Reactions have instead a peaked distribution of connections, the peak being located at the average degree of all reactions $\langle k_R \rangle$. This property arises from the fact that reactions have a bounded number of participants, typically from 2 to 12. The most typical case is when reactions have 4 metabolites, leading to a peaked distribution with the maximum around $k_R = 4$.

• Another common feature is the *small-world* property,[17] meaning that all nodes in the network can be reached within a few jumps or hops following connections in the network. In terms of the average shortest path length d, which is the average of all the shortest distances between pairs of nodes,

$$d \propto \ln N \qquad (2)$$

where N is the number of nodes. For metabolic networks, the authors in Ref. 17 computed the average path length in the one mode projection of metabolites and extracted a value of ~ 3.

• Clustering is a measure of the triadic closure of nodes in the network.[18,19] The clustering coefficient is the probability that three nodes are connected to each other. It gives a measure of the number of triangles in a network as a function of the degrees of the nodes. Usually, in metabolic networks, nodes tend to aggregate in groups, displaying high levels of clustering.[20]

1.2. *Function: Flux Balance Analysis*

On what follows, we compute activities and essentialities of reactions. To do so, we need to compute fluxes in the metabolic network. Constraint-based optimization techniques are very useful to obtain them without using kinetic parameters.[21,22] In particular, the technique called Flux Balance Analysis (FBA)[23] has been proved very successful in order to predict growth rate values or gene lethality in metabolic networks[24,25] (see Fig. 2).

The idea of FBA is simple. It starts from the typical kinetic equations which describe the temporal variation of a metabolite i, derived from the mass conservation principle,

$$\frac{dc_i}{dt} = \sum_{j=1}^{N_R} \alpha_{i,j}\nu_j, \qquad (3)$$

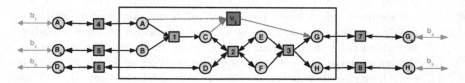

Fig. 2.   Example of a FBA calculation in a metabolic network. Reactions are denoted by squares and metabolites by circles. Note that exchange reactions are labeled with b_x in order to differentiate them from metabolic reactions, as well as the biomass production reaction, which is labeled as ν_g.

where $\alpha_{i,j}$ is the stoichiometric coefficient of metabolite i in reaction j, ν_j is the flux of reaction j and N_R is the total number of reactions in the network. Since a metabolic network is an open-system, some metabolites will leave or enter the organism. Thus, it is not possible to arrive to a thermodynamic equilibrium state. However, it is assumed that the system will reach a stationary state compatible with the external constraints. This condition turns Eq. (3) into Eq. (4)

$$0 = \sum_{j=1}^{N_R} \alpha_{i,j}\nu_j, \tag{4}$$

viewed as a product of a stoichiometric matrix by a vector of fluxes (Eq. (5)),

$$0 = S \cdot \nu. \tag{5}$$

Constraints $\alpha \leq \nu_j \leq \beta$ must be imposed to bound the values of the fluxes, where α and β are values chosen to render the whole scheme realistic both chemically and biologically. In addition, to simulate the environment, exchange fluxes are imposed for metabolites exchanged with the environment (see Fig. 2).

In metabolic networks, there are usually more reactions than metabolites, leading to multiple solutions and Eq. (5) being underdetermined. Thus, a biological objective function is introduced to restrict the possible solutions of the system and to obtain only a particular biologically relevant one. This objective function depends on the biological information that one wants to extract, but usually one chooses to optimize biomass formation and therefore to maximize growth. Other possible objective functions are ATP or NADH production. To summarize, we have a system of equations and we try to find a solution that optimizes the value of an objective function. When choosing biomass production, if no solution exists we assume that the system is not able to grow and therefore we conclude that the organism is not able to survive.

Apart from the definition of the objective function, other constraints are used in order to get reliable results. Thermodynamic constraints like reversibility of reactions are added in order to constrain the possible solutions. The exchanges with the environment are set in terms of auxiliary fluxes which simulate the consumption and expulsion of metabolites in the environment. Notice in passing that FBA results will thus depend on the particular chosen medium.

To calculate the effect of knockouts of reactions, the selected reaction is removed from the network which is equivalent to force the inability of the network to use the selected reaction. As a consequence, the system can respond in three different ways:

(1) The growth rate is unaltered.
(2) The growth rate is decreased, which means that the biomass formation of the organism is reduced but the organism is still alive.
(3) The growth rate takes a null value, meaning that the performed knockout is lethal for the organism.

1.3. *Previous work on activity and essentiality of reactions* in silico

Since we will make an extended use of the concepts of activity and essentiality of a particular reaction of a metabolic network in the following sections, it is useful to describe previous work referring to both concepts. Briefly, activity refers to the case when a reaction has a nonzero flux, whereas essentiality refers to the case when the removal of the chosen reaction leads to supression of growth for the whole organism.

Although activity and essentiality have been referred to genes, essentiality can be extended to reactions, since a gene regulates the production of the enzyme which catalyzes the reaction to be knocked out. In this way, validations of the proposed reconstructions of organisms were done in the past.[26-28]

Closer to our approach entirely based on reactions, the notions of activity and essentiality have been discussed in recent works with the purpose of extracting metabolic cores.[12,13] More precisely, the authors in Ref. 12 identified the metabolic core as the subset of always active reactions of *E. coli* for a set of media which differ in the composition of nutrients. This analysis is repeated for two additional organisms, *H. pylori* and *S. cerevisiae*, the conclusion being that major portions of the metabolic core have been conserved through evolution.

Wagner and coworkers[13] took another approach by studying quantitatively the extent to which reactions cannot be bypassed in a metabolic network, a property called superessentiality. Bypassing happens when an organism is able to activate other reactions in order to compensate the effect of the removal of a reaction. The methodology to define this superessentiality property first proceeds by taking the biomass composition of *E. coli* and adding all the known biochemical reactions to define a universal network. For this network, the essentiality of all reactions is computed with FBA and in this way the authors formed a core of 133 reactions which cannot be bypassed, i.e., they are super-essential. Alternatively, the authors propose a complementary way to compute superessentiality. To do so, the authors generated a large ensemble of randomly extended and viable networks, by starting from the metabolic network of *E. coli* and adding new reactions at random. For each of these realizations, the authors computed the essentiality of the participating reactions, and then calculated a super-essentiality index as the fraction of these networks where a particular reaction is essential. In this way, Wagner and coworkers extracted 139 super-essential reactions that contained the core of 133 reactions mentioned above. Finally, by considering different media, the authors concluded that super-essentiality is not very sensitive to the environment whereas single essentiality does depend on the considered medium.

2. Methods

2.1. *Genome-scale representation of metabolism*

We work with the *i*JO1366 reconstruction of the K-12 MG1655 strain of *E. coli* given in Ref. 29.

The metabolic network reconstruction of *E. coli* contains 2250 biochemical reactions, 333 auxiliary reactions (exchange, biomass, sink and ATP maintenance reactions) and 1805 metabolites. Auxiliary reactions are reactions which are not in the real metabolic network of the organisms but that are necessary to compute FBA flux solutions. These auxiliary reactions may correspond to

- Exchange reactions as explained in Sec. 1.2.
- The biomass reaction is a reaction which simulates the biomass production and thus the growth of the organism.
- Sink reactions are necessary when reactions which consume a metabolite have not been identified. This reaction has the simple form $A \rightarrow$, and has the utility of simulating the consumption of a metabolite.
- The ATP maintenance reaction is a reaction which consumes ATP in order to simulate other biological energetic costs for the organism which are not associated to growth.

Since we have two entities, metabolites and reactions, we use a bipartite version to represent the network. Note also that metabolites in different compartments are treated as different metabolites, and transport reactions are used to connect them.

2.2. *Media composition*

We perform FBA calculations on the set of minimal media given in Ref. 29, which consist of a set of mineral salts, and one source of carbon, nitrogen, sulfur and phosphorus. A minimal medium is the minimal set of metabolites which ensure the viability of an organism. To construct them, a set of minerals salts and four extra metabolite families are used, each of these four metabolites representing carbon, nitrogen, phosphorus and sulfur sources respectively. The set of mineral salts is always the same, each source family is varied while the other three sources are fixed to the standard metabolites of each kind (C: glucose, N: ammonium, P: phosphate, S: sulfate). A total number of $n^0_{media} = 555$ can be constructed in this way, with a final $n_{media} = 333$ which allow growth after computing the organism growth rate.

In addition, we construct 10000 random media, of which 3707 give a nonzero growth, by allowing the 90% of metabolites which can act as nutrients. Each random medium contains at least the altogether minimal salts and the representatives of the four families of basic elements.

2.3. *FBA implementation*

The software that we use to perform the calculations of FBA is GLPK, through its associated solver GLPSolver. We use a dual *simplex* algorithm to compute the solutions. It is a variant of the normal simplex algorithm. The latter is an iterative algorithm which is based on finding first a feasible solution and then finding the most optimal solution based on the feasible solution. On the contrary, dual simplex works by first finding an optimal solution and then finding a feasible solution, again, if it exists.

2.4. *Quantitative definition of activity and essentiality*

The activity $a_{i,j}$ of a reaction i in a medium j is defined as

$$a_{i,j} \equiv \begin{cases} 1 & \text{if } \nu_{i,j} > 0 \\ 0 & \text{if } \nu_{i,j} = 0 \end{cases} \tag{6}$$

where $\nu_{i,j}$ denotes the flux of reaction i in medium j.

To obtain a significant value of the activity, we choose to perform FBA calculations for the minimal and random media. In addition, the activity is normalized by dividing it for the number of media in which we have performed the calculation of the fluxes. Therefore, the activity of a reaction i for a given set of media n_{media} will be obtained according to

$$a_i \equiv \frac{1}{n_{media}} \sum_{j=1}^{n_{media}} a_{i,j} \tag{7}$$

with $0 \le a_i \le 1$.

Essentiality is defined on the subset of active reactions. To discern the essentiality of a particular reaction we look for the FBA growth rate after constraining the bounds of the selected reaction to zero. An expression of the essentiality of a reaction i in a medium j is given as

$$e_{i,j} \equiv \begin{cases} 0 & \text{if } \nu_{g,j} > 0 \\ 1 & \text{if } \nu_{g,j} = 0 \end{cases} \tag{8}$$

where $\nu_{g,j}$ is the FBA flux of reaction of production of biomass in medium j. Again, we average the results on several media and we normalize essentiality by dividing by the number of media. In this way, the bounds of essentiality of a reaction will lay between 0 and the corresponding activity of the reaction, $e_i \le a_i$,

$$e_i \equiv \frac{1}{n_{\text{media}}} \sum_{j=1}^{n_{\text{media}}} e_{i,j}. \tag{9}$$

Another useful magnitude is the fraction of media where reaction i is essential over the number of media where it is active. This is computed with $p_i = \frac{e_i}{a_i}$.

3. Results and Discussion

In this section, we evaluate activities and essentialities of reactions in *E. coli* as explained before. After computing FBA on all media and all mutants, we plot essentiality versus activity for all reactions of both sets of media, and we expect a representation where all the points must fall on the diagonal or under it. This representation is shown in Fig. 3 for minimal and random media.

From the representation given in Fig 3, one can classify reactions into several categories:

(1) **Essential whenever active reactions:** $0 < a_i = e_i$. They are essential in all media where they are active. These reactions lay on the diagonal.

(2) **Always active reactions:** $a_i = 1$, $0 < e_i < a_i$. They are always active and sometimes essential. These reactions are located in the opposite y axis.

(3) **Never essential reactions:** $0 < a_i < 1$, $e_i = 0$. They are never essential but sometimes active. These reaction are located in the x axis.

(4) **Partially essential reactions:** $0 < a_i < 1$, $0 < e_i < a_i$. They are essential a fraction of times when they are active. These reactions are located inside the triangle formed by the diagonal, y and x axes.

To study the obtained results, we will focus on the different subnetworks, obtained by filtering the complete original network according to

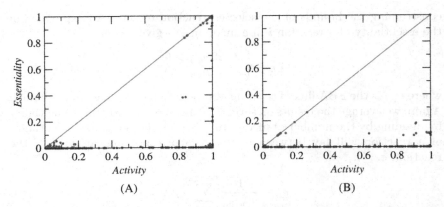

Fig. 3.   Representation of essentiality versus activity. (A) Minimal media. (B) Random media. In both pictures the four different categories can be clearly differentiated. Diagonal: *essential whenever active reactions*. Opposite y axis: *always active reactions*. x axis: *never essential reactions*. Inside triangle: *partially essential reactions*.

the four basic explained categories. More precisely, these sub-networks are obtained by only maintaining in the network reactions within the respective mentioned categories. Once we have these sub-networks, we compute the number of connected components (CC) in the network in order to know whether the selected network is fragmented or not. This in turn allows us to compute the giant component (GC) and the strongly connected component (SCC)[30] of the networks. The giant component is the largest connected component of the network, whereas the SCC is a portion of the giant component where for every two nodes one can always find a directed path connecting them.[31] This study in terms of the connected components is done in order to detect whether reactions of each type are responsible of the percolation of the network or to know whether a kind of reaction is predominant over the others, depending on the number of reactions present in the sub-network.

In Table 1 we summarize the statistics of active and essential reactions and we show the statistics of the large scale components of the sub-networks they span. A precise discussion of such statistics is provided on what follows. Notice also in passing that there are several reactions which are strictly never active (902). This may be explained by the fact that we are working with minimal media. In addition, we can see that the complete network, which correspond to values of activity of $0 \leq a_i \leq 1$ and essentiality of $0 \leq e_i \leq a_i$, is constituted by a single giant connected component and that, in addition, it has a large strongly connected component, typical of metabolic networks.

Table 1. Connected components and number of reactions N_R in each sub-network. Values in parentheses correspond to percentages. GC percentages are computed by dividing its absolute value by Complete, whereas SCC is computed by dividing it by GC. Categories in bold correspond to the four basic categories identified in the text.

Category	CC	N_R
Essential whenever active reactions	**Complete**	**665**
$a_i > 0$	**GC**	**611(91.9)**
$e_i = a_i$	**SCC**	**409(66.4)**
Always active reactions	**Complete**	**37**
$a_i = 1$	**GC**	**34(91.9)**
$0 < e_i < a_i$	**SCC**	**29(85.3)**
Never essential reactions	**Complete**	**494**
$0 < a_i < 1$	**GC**	**494**
$e_i = 0$	**SCC**	**476(96.4)**
Partially essential reactions	**Complete**	**152**
$0 < a_i < 1$	**GC**	**145(95.4)**
$0 < e_i < a_i$	**SCC**	**129(90.0)**
All reactions	Complete	2250
$0 \leq a_i \leq 1$	GC	2250
$0 \leq e_i \leq a_i$	SCC	2076(92.0)
Never active		
$a_i = 0$	Complete	902
$e_i = 0$		
Essential and active in some media	Complete	458
$0 < a_i < 1$	GC	409(89.3)
$e_i = a_i$	SCC	200(48.8)
Essential and active in all media	Complete	207
$a_i = 1$	GC	198(95.7)
$e_i = a_i$	SCC	174(86.1)

3.1. *Essential whenever active reactions*

A histogram of the values of the essentiality of the set of reactions *essential whenever active reactions* is shown in Figs. 4A and 4B. A bimodal distribution is clearly displayed, with a peak at $a_i = e_i \simeq 0$ and the other at $a_i = e_i \simeq 1$. This means that there is a core of reactions that are always active and essential, and there is another set of reactions that are active a very few times, but all these times are essential. This histogram coincides with the classification of the dependence of essentiality on the environment given in Ref. 13. The peak at values of activity ~ 0 correspond to *environment-specific* essential reactions, whereas the peak at

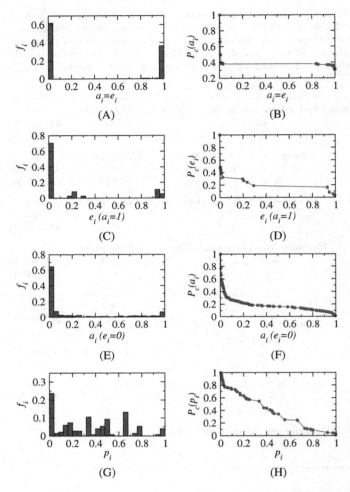

Fig. 4. Histograms (fraction) and complementary cumulative probability distribution function of activity/essentiality (depending on the category) for minimal media. (A), (B) Essential whenever active reactions. (C), (D) Always active reactions. (E), (F) Never essential reactions. (G), (H) Partially essential reactions.

values of activity ~ 1 correspond to *environment-general* essential reactions. The first includes reactions whose deletion abolishes growth in a very few environments, whereas the second ones correspond to reactions whose deletion suppresses growth in all environments.

A deeper characterization of this set of reactions is made in Table 1. We apply to this network the same algorithm to compute the GC and the

SCC. In Table 1 we show that the subnetwork of *essential whenever active reactions* has a large giant component, nearly 90% of the sub-network. If the reactions with activity-essentiality index of 1 are excluded from this sub-network, we obtain another subset which has also a large GC (89.3% of the total 458 reactions). This means that reactions with $a_i = e_i = 1$ are not responsible for the percolation state of the sub-network of *essential whenever active reactions* and points out to a large degree of redundancy.

An illustrative example of this particular category of reaction is *Potassium transport (Ktex)*. It is a reaction which supplies the organism with potassium. Potassium is an important metabolite which influences the osmotic pressure through the cell membrane and it also secures the propagation of electric impulses. Since they are important processes of organisms, this reaction is always active in order to secure that they are done properly. As a result, this reaction must always be active in order to have a nonzero growth.

For random media (see Figs. 5A and 5B), a similar behavior is obtained, with larger probabilities at the extrema, but an extra peak is obtained for low values of activity-essentiality, meaning that there are some reactions which are not as specific as *environment-specific* reactions because they are active and essential in more than one medium, losing in this way their specificity. This makes sense for random media, since they contain many metabolites that may activate many reactions and in this way they lose the specificity of a minimal medium, which actives only the reactions that allow an organism to grow.

3.2. *Always active reactions*

The set of reactions called *Always active reactions* contains reactions with $a_i = 1$ and $e_i \leq a_i$. In Figs. 4C and 4D we see that in this case we have a large peak at values of $e_i = 0$, meaning that the largest part of reactions with $a_i = 1$ have a value of $e_i = 0$. This means that, although some reactions are always active, they are not essential. One may be tempted to think that reactions with very low values of essentiality are then useless and hence they should be removed from the network. Nevertheless, there are two reasons that justify their consideration.

- The first one is that these reactions may improve the life conditions of the organism. These reactions, in spite of being non-essential, might be active in order to increase the growth of the organism. In order to survive to hard conditions, an organism which is able to reproduce fast and efficiently will, with large probability, survive to unfriendly life conditions. Especially, these reactions which are always active but never essential might be important growth magnifiers.

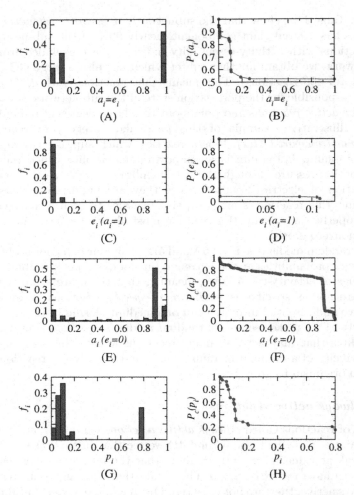

Fig. 5. Histograms (fraction) and complementary cumulative probability distribution function of activity/essentiality (depending on the category) for random media. (A), (B) Essential whenever active reactions. (C), (D) Always active reactions. (E), (F) Never essential reactions. (G), (H) Partially essential reactions.

- The second one is more subtle. The removal of these reactions could form a synthetic lethal pair. Synthetic lethality is a concept based on the idea that the simultaneous deletion of two reactions, which are not essential separately, is lethal.[32] As an example, two reactions regulated by the genes *tktA* and *tktB*, which are in the peak at $e_i = 0$ and $a_i = 1$, form a synthetic lethal pair,[33] and the removal of these reactions would abolish growth by impeding the synthesis of nucleotides, nucleic acids and aromatic amino acids. We have exhaustively considered this question in a recent separated publication.[34] Briefly, the reactions regulated by these genes are *TKT1* and *TKT2*, both with a complete name of *Transketolase*, reactions which belong to the *Pentose Phosphate Pathway* pathway. This pathway generates NADPH and pentoses phosphate, the latter being a precursor used in the synthesis of nucleotides, nucleic acids and aromatic amino acids. Both reactions are always active to ensure a large production of these mentioned products, and when one of these reactions is knocked out, the other reaction is in charge to restore this function.

In Table 1 one can see that as in *essential whenever active reactions*, this category of reactions form a network with a giant connected component which is almost the full network with also a large SCC.

Note that for random media (see Figs. 5C and 5D), a similar trend is obtained compared to the minimal media.

3.3. *Never essential reactions*

Never essential reactions have values of activity and essentiality which satisfy $e_i = 0$ and $0 < a_i < 1$. The histogram of the values of the activities for these reactions is shown in Figs. 4E and 4F. We recover again a similar histogram to that corresponding to *always active reactions*. This means that, not surprisingly, the largest part of non-essential reactions are not much active. The individual removal of these reactions will leave the growth rate unaltered or reduced. The existence of these reactions could be explained again in terms of improving the growth of the organism.

An example of a reaction of this kind is *Manganese transport in via permease (no H+)*, *MN2tpp*, a reaction which pumps manganese into the organism. Its non-essentiality comes from the fact that there exists an alternative reaction called *Manganese (Mn+2) transport in via proton symport (periplasm)*, *MNt2pp*, which also pumps manganese into the organism, but the latter uses a proton gradient to perform the transport.

In Table 1 one can see again the same trend, there is a giant connected component that is almost the full network with a large SCC.

For random media, different results are obtained in this case (see Figs. 5E and 5F). The largest peak is located at large values of activity, which means that there is a large set of reactions which are active but not essential. The peak located above 0.8 could appear due to the fact that these random media that we have used are in fact rich media, since there are a lot of available nutrients in the environment. Thus, it is possible that a common set of metabolites activates the same reactions for many media, resulting in the mentioned peak. These reactions are responsible to increase the value of the flux of the biomass reaction.

3.4. *Partially essential reactions*

Partially essential reactions contains reactions with activities with values of $0 < a_i < 1$ and $0 < e_i < a_i$. Since these reactions have values of essentiality and activity different from 0, we choose to represent the histogram in terms of $\frac{e_i}{a_i}$, as shown in Fig. 4, panels G and H. The distribution is rather homogeneous, meaning that, despite the fact that one reaction must be active as a previous step for being essential, this does not imply that all active reactions are essential, meaning that there is correlation in the sense that essentiality implies activity but activity does not always imply essentiality.

Table 1 shows again a large giant connected component containing a large SCC. Notice that this trend has been maintained for all categories of reactions, meaning that these categories form networks of reactions where it is possible to arrive from one reaction to a large part of the reactions of the network.

Again, different results are obtained for random media (see Figs. 5G and 5H). In this case, we do not obtain homogeneously distributed values as in minimal media, instead they are concentrated at low values and at a value of $\frac{e_i}{a_i} = 0.8$. This means that reactions are essential in fewer environments where they are active, showing again that activity does not imply essentiality.

4. Conclusions

The study of activity and essentiality of reactions is crucial in order to know how the metabolism of organisms behave in a given environment and its fitting capacities to different media. In this work, we quantify both magnitudes and show how they vary in a set of environments. We have classified both magnitudes depending on the behavior they show across different media. In particular, we have identified four categories which depend on the values of essentiality and activity.

With this study we have recovered the set of essential reactions given in Ref. 13 respectively called *environment-specific* and *environment-general* reactions. They correspond to the bimodal behavior in our category called *essential whenever active reactions*. Given their importance, these reactions can be selected as drug targets since they are fundamental constituents of the metabolism of *E. coli* organism. Another important feature that we have observed is the fact that some reactions, in spite of being nonessential, are active in order to increase the growth rate of the organism and to prevent the formation of synthetic lethal entities. The categories of reactions which show this behavior are *always active reactions* and *never essential reactions*. The last feature that we can extract from the last category, *partially essential reactions*, is that active reactions does not imply to be also essential. Therefore, extrapolating activity to essentiality is not correct for these reactions.

This study allows to better understand the metabolism of *E. coli* by means of flux studies across different environmental conditions. In summary, we have extended the concept of core of reactions by computing the essentiality of reactions in addition to the activity. This is an important point, since we have detected that not all reactions which are active in all media are essential, which also supports the already mentioned idea about the lack of correlation between activity and essentiality, noting again that a reaction must be active to be essential but it is not a unique requirement. This distinction between active and essential reactions allows us to separate reactions between different categories, each one displaying different properties, which allows to better understand the metabolism of *E. coli.*

Acknowledgments

This work was supported by MICINN Projects No. FIS2006-03525, FIS2010-21924-C02-01 and BFU2010-21847-C02-02; Generalitat de Catalunya grant No. 2009SGR1055; a 2013 James S. McDonnell Foundation Scholar Award in Complex Systems, the Ramón y Cajal program of the Spanish Ministry of Science, and the FPU grant of the Spanish Ministry of Science.

References

1. H. Jeong, B. Tombor, R. Albert, Z. N. Oltvai and A. L. Barabási, *Nature* **407**, 651–654 (2000).
2. H. W. Ma and A. P. Zeng, *Bioinformatics* **19**, 270–277 (2003).
3. N. D. Price, J. L. Reed and B. Ø. Palsson, *Nat. Rev. Microbiol.* **2**, 886–897 (2004).
4. R. Albert and A. L. Barabási, *Rev. Mod. Phys.* **74**, 47–97 (2002).
5. M. E. J. Newman, *SIAM Rev.* **45**, 167–256 (2003).
6. V. Hatzimanikatis, C. Li, J. A. Ionita, C. S. Henry, M. D. Jankowski and L. J. Broadbelt, *Bioinformatics* **21**, 1603–1609 (2004).
7. S. N. Dorogovtsev, A. V. Goltsev and J. F. F. Mendes, *Rev. Mod. Phys.* **80**, 1275–1335 (2008).

8. R. Albert, H. Jeong and A. L. Barabási, *Nature* **406**, 378–382 (2000).
9. A. G. Smart, L. A. N. Amaral and J. Ottino, *Proc. Natl. Acad. Sci. USA* **105**, 13223–13228 (2008).
10. O. Güell, F. Sagués, G. Basler, Z. Nikoloski and M. Á. Serrano, *Journal of Computational Interdisciplinary Sciences* **3**(1–2), 45–53 (2012).
11. O. Güell, F. Sagués and M. Á. Serrano, *Sci. Rep.* **2**, 621 (2012).
12. E. Almaas, Z. N. Oltvai and A. L. Barabási, *PLoS Comput. Biol.* **1**, 0557–0563 (2005).
13. A. Barve, J. F. M. Rodrigues and A. Wagner, *Proc. Natl. Acad. Sci. USA* **1091**, E1121–E1130 (2012).
14. B. Ø. Palsson, *System Biology: Properties of Reconstructed Networks* (Cambridge University Press, Cambridge, 2006).
15. R. Guimerà and L. A. N. Amaral, *Nature* **433**, 895–900 (2005).
16. M. Boguñá and M. Á. Serrano, *Phys. Rev. E* **72**, 016106 (2005).
17. D. A. Fell and A. Wagner, *Nat. Biotechnol.* **18**, 1221–1122 (2000).
18. M. Á. Serrano and M. Boguñá, *Phys. Rev. E* **74**, 056114 (2006).
19. M. Á. Serrano and M. Boguñá, *Phys. Rev. E* **74**, 056115 (2006).
20. E. Ravasz, A. L. Somera, D. A. Mongru, Z. N. Oltvai and A. L. Barabaási, *Science* **297**, 1551–1555 (2002).
21. C. H. Schilling and B. Ø. Palsson, *Proc. Natl. Acad. Sci. USA* **95**, 4193–4198 (1998).
22. C. H. Schilling, J. S. Edwards, D. Letscher and B. Ø. Palsson, *Biotechnol. Bioeng.* **71**, 286–306 (2000).
23. J. D. Orth, I. Thiele and B. Ø. Palsson, *Nat. Biotechnol.* **28**, 245–248 (2010).
24. A. M. Feist et al., *Mol. Syst. Biol.* **3**, 121 (2007).
25. A. R. Joyce and B. Ø. Palsson, *Methods Mol. Bio.* **416**, 433–457 (2008).
26. J. S. Edwards, R. U. Ibarra and B. Ø. Palsson, *Proc. Natl. Acad. Sci. USA* **97**, 5528–5533 (2000).
27. J. S. Edwards and B. Ø. Palsson, *Nat. Biotechnol.* **19**, 125–130 (2001).
28. P. F. Suthers, M. S. Dasika, V. S. Kumar, G. Denisov, J. I. Glass and C. D. Maranas, *PLoS Comput. Biol.* **5**, e1000285 (2009).
29. J. D. Orth et al., *Mol. Syst. Biol.* **7**, p. 535 (2011).
30. H. W. Ma and A. P. Zeng, *Bioinformatics* **19**, 1423–1430 (2003).
31. A. Broder et al., *Comput. Netw.* **33**, 309–320 (2000).
32. P. Novick, B. C. Osmond and D. Botstein, *Genetics* **121**, 659–674 (1989).
33. G. Zhao and M. E. Winkler, *J. Bacteriol.* **176**, 883–891 (1995).
34. O. Güell, F. Sagués and M. Á. Serrano, *PLos Comput. Biol.* **10**(5), e1003637 (2014).

Chapter 4

CHEMICALLY-DRIVEN BIOLOGICAL BROWNIAN MACHINE

Mitsuhiro Iwaki

Quantitative Biology Center, RIKEN,
6-2-3 Furuedai, Suita, Osaka 565-0874, Japan

Graduate School of Frontier Biosciences, Osaka University,
1-3 Yamadaoka, Suita, Osaka 565-0871, Japan

1. Introduction

Motor proteins are natural molecular machines that convert the chemical free energy of ATP into mechanical work. Myosin is a motor protein that is originally famous for being the functional unit of muscle contraction, but has since been found to describe a diverse superfamily of molecular machines.[1] Due to certain ideal features suitable for single molecule study, myosin-V and -VI are particularly well characterized from the viewpoint of single molecule mechanical properties, atomic structure, enzymatic properties and physiological functions.

Both myosin-V and -VI are dimers and function as vesicle transporters in cells.[2] Although their structural details are different, both are composed of the motor domain (traditionally called "head") and the rod-like domain, the lever-arm, which is 10–30 nm in length (Fig. 1). The head interacts with cytoskeletal actin filament and hydrolyses ATP. These two domains are the minimal unit for myosin function.[3]

2. Dynamics of Single Myosin Motor Proteins

Single molecule imaging and manipulation techniques have been groundbreaking in understanding the molecular mechanism of motor function.[4] One fundamental method is total internal reflection fluorescence microscopy

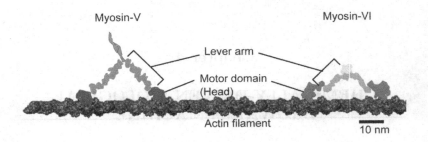

Fig. 1.   Structure of Myosin-V and -VI bound to an actin filament.

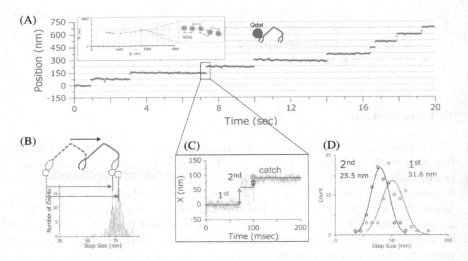

Fig. 2.   Single molecule imaging of myosin-V. (A) Trajectory of a quantum dot (Qdot) attached to a myosin-V head with 33 ms time resolution. Inset: Mapping of the Qdot position on the imaging plane. (B) Histogram of step size. (C) High-speed imaging of the internal dynamics of the step. (D) Histogram of the 1st and 2nd sub-step size.

(TIRFM),[5] which can visualize the motion of single molecules via observation of an attached single fluorophore. We have adopted quantum dots (semiconductor particles of 10 nm size) as our bright fluorescent probe and attached them specifically to the head of myosin-V or -VI molecules using protein engineering to observe motor dynamics with 1 nm accuracy (Fig. 2).

Figure 2A shows the spatial and temporal trajectory of a quantum dot attached to a myosin-V head moving along an actin filament as observed by TIRF. The head rapidly and repeatedly moved in one direction within the frame rate of the imaging camera (33 ms). The observed step size (76 nm on average) is consistent with a hand-over-hand mechanism where the two heads alternately move forward in a manner similar to walking (Fig. 2B). The 76 nm step size distributed as two peaks, 72 nm and 78 nm, suggesting the myosin head precisely catches the 13th or 14th forward actin monomer relative to the detachment position from the filament.

To clarify the mechanism that decides the directionality and step size, we constructed a high-speed imaging system to visualize the internal dynamics of the steps. For this purpose, we attached a gold nano particle (40 nm in diameter) to the myosin head instead of the quantum dot and monitored the scattered light, which is much brighter than the fluorescence. Thus, we could monitor the image at a time resolution of 37 μs (∼1000-fold more rapid than quantum dot imaging).[6,7] At this timescale, the 76 nm steps could be divided into 51 nm and 25 nm sub-steps (Figs. 2C and 2D). The 51 nm sub-steps occurred rapidly (<100 μs), after that, we could see Brownian motion with an average dwell time of 16 ms, then the Brownian head caught forward actin. The 51 nm sub-step is consistent with lever-arm swinging by the front head, a phenomenon directly observed by Shiroguchi *et al.*[8] The two-sub-step motion has led to a model where myosin moves by a lever-arm swing (structural change) and Brownian search-and-catch. While this model can explain why myosin moves forward during the lever-arm swing, it cannot explain why myosin does so during the Brownian seach-and-catch. One theory is presented in Sec. 4.

3. Mechano-Chemical Coupling of Myosin-V and -VI

Myosin dynamics should be coupled with the ATP hydrolysis cycle. Figure 3 shows the mechano-chemical coupling model of myosin-V and -VI obtained from the transient kinetics of the ATP hydrolysis cycle by biochemical assays[9] and single molecule experimental data.[10] The rate limiting state is the two-headed bound state with the heads bound to adenosine diphosphate (ADP), a product of hydrolyzed ATP, or no nucleotide (Fig. 3, state 1). After the rear head releases ADP, it is bound by ATP and detaches from actin by a conformational change (state 2). Then, the front head makes a lever-arm swing, and the detached head undergoes Brownian motion back and forth (Brownian search) (state 3). During the Brownian search, the head rapidly hydrolyzes the bound ATP into ADP and Pi (inorganic phosphate). During this time, it is possible that the Brownian head weakly interacts with the many accessible myosin binding sites on the actin filament. When Pi is released from the myosin head, the actin binding becomes strong and the forward step is completed (state 4).

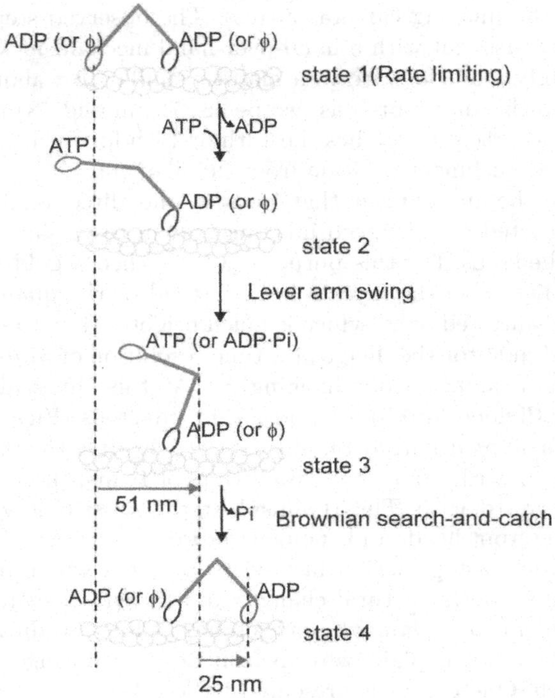

Fig. 3. Mechano-chemical coupling of myosin-V hand-over-hand motion.

Uemura and Ishiwata first reported that the rate constant of ATP hydrolysis by kinesin, another motor protein that shares many similarities with myosin, is mechano-sensitive such that the direction of mechanical strain can accelerate or suppress the rate.[11] Several groups have since examined this mechano-sensitivity in myosin-V and -VI[12–14] (a summary is given in Table 1). The results can well explain the hand-over-hand mechanism. Intramolecular strain should occur when myosin-V or -VI forms a two-headed bound state.[13] Thus, the rear head senses a forward strain (forward direction is defined as the direction of myosin movement) and the front head senses a backward strain. Because the forward strain accelerates ADP release and/or ATP binding, detachment of the rear head from actin should be accelerated. In contrast, because backward strain suppresses ADP release and/or ATP binding, the detachment of the front head should be suppressed. As a result, myosin-V and -VI coordinate their two heads by predominantly detaching their rear head, which results in hand-over-hand motion.

Table 1. Mechano-sensitivity of the transient kinetics of ATP hydrolysis.

	Myosin-V		Myosin-VI	
	Forward strain	Backward strain	Forward strain	Backward strain
ADP release	Accelerated	Suppressed	Slightly accelerated	None
ADP binding	Suppressed	Suppressed	Accelerated	Accelerated
ATP binding	None	None	Accelerated	Suppressed

4. Strain-Sensor Mechanism

The reported mechano-sensitivity of the transient kinetics of the ATP hydrolysis cycle (i.e. ADP release/binding and ATP binding) is analyzed for the rate-limiting steps in the cycle.[12−14] Analysis for non-rate-limiting steps such as Pi release was difficult due to the rapid transition. When the Brownian head is bound to both ADP and Pi, it undergoes weak interactions with actin, which can be described as an equilibrium state between rapid (sub-millisecond) attachments and detachments.[15] Our team have constructed a monitoring system that can directly observe the weak binding state by detecting weak attachments between myosin and actin.[16] We found that Pi release, which corresponds with the transition from weak to strong binding, was very mechano-sensitive. This strain-sensor mechanism, then, can explain the decision-making that enables the forward catch by the Brownian head, which is described in detail in this section.

4.1. Quantification of mechano-sensitivity for the weak-to-strong transition

To monitor the weak attachment of the Brownian head with actin, fluorescent polystyrene beads tagged to myosin-VI heads (myosin beads) were rapidly scanned along a single actin filament bridged onto a glass slide using optical tweezers (Fig. 4A). If the myosin bead is positioned onto an actin filament in a stationary manner, the short-lived (sub-milliseconds) weak binding events should be hidden in the Brownian motion of the myosin bead and undetectable. However, we successfully visualized the binding by applying optical tweezers to single myosin beads. We scanned the beads in both directions of the actin filament at a scanning speed of up to $4\,nm/\mu s$ (or loading rate of $0.26\,pN/\mu s$).

The expected transition of a weakly bound head tethered to an optically-trapped bead when scanned is shown in Fig. 4B. The head rapidly alternates between attachments and detachments in the weak-binding state. If the bead is scanned while the head is attached to actin (3), the bead displaces until the tether is taut and then stops (4). When the weakly

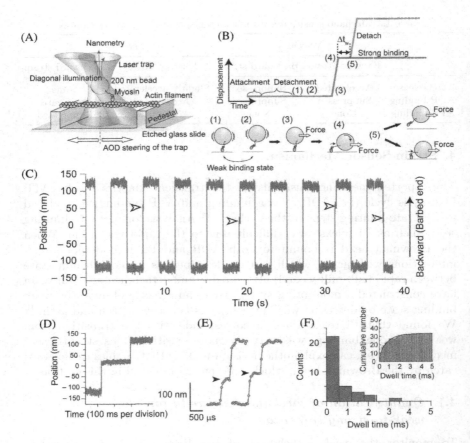

Fig. 4.   High-speed scan of a myosin-VI bead along an actin bridge. (A) Experimental geometry. (B) Expected trajectory. (C) Typical experimental trajectory. Arrowheads indicate binding events during the scan. Long-lived strong binding (D) and short-lived weak binding (E) during scan. (F) Dwell time distribution for weak binding.

bound head is strained for a moment (Δt_w), the head transitions to strong binding or detaches (5). We measured this transition while applying forward or backward strain to the head (Fig. 4C).

We frequently observed long-lived, strong bindings during scans (Fig. 4D). The distribution of the dwell times of strong bindings at the maximum backward loading rate (0.26 pN/μs) was expressed as a two-step reaction with dwell times of 116 ms and 8.1 ms. These values are consistent with ADP release and ATP waiting times during the strong binding state, respectively, as previously observed under loaded conditions.[17]

Also at the maximum loading rate, we observed short-lived attachments (Figs. 4E and 4F). The dwell time distribution fit to a single exponential with time constants of $310\,\mu s$ and $1.9\,ms$ during backward and forward scans, respectively. This time range is hundredfolds shorter than that of strong binding ($116\,ms$). Therefore, the short-lived attachments are likely due to weakly bound heads that failed to transitions to strong binding and instead detached from actin while being strained (Fig. 4B).

Strong bindings were more frequently observed for backward (barbed end of actin filaments for myosin-VI) scans than forward scans at the maximum loading rate (Fig. 5A). When the loading rate was decreased to $0.003\,pN/\mu s$, the frequency of strong bindings during backward scans decreased such that it became equal to that of forward scans. Furthermore, the frequency of strong bindings during forward scans was independent of the loading rate. It is thought that weakly bound heads were fully strained at the maximum loading rate, but only partially strained at low loading rates. Therefore, the results strongly suggest that the transition of weak binding to strong binding is accelerated by backward strain, but independent of forward strain.

Strong binding during forward scans is likely the result of the head having already formed a strong bond with actin before the scan. Therefore,

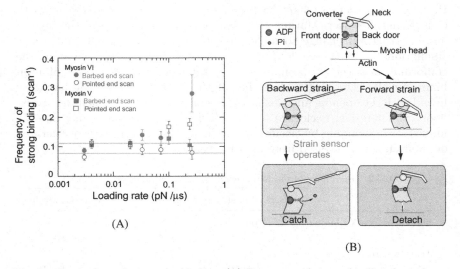

(A)

(B)

Fig. 5.   Strain-dependent strong binding. (A) Frequency of strong binding depends on loading rate. (B) A model describing how strong binding is accelerated in a strain depending manner (see text for details).

the frequency of strong bindings during forward scans was independent of the scanning speed. Such strong bindings should occur during backward scanning as well, such that the net strong bindings formed during backward scanning should be the difference between strong bindings during backward and forward scans. We found that the net frequency of strong bindings when strained backward at the maximum loading rate was ~0.2 scan^{-1}.

The frequency of weak binding to detachment events at the maximum backward loading rate was 0.031 ± 0.019 scan^{-1}, which was about 6-fold smaller than that of strong bindings (0.2 scan^{-1}). The results show that when rapidly strained backward, 86% of weakly bound heads transition to strong binding and the remaining 14% detach from actin over an average time of $310\,\mu$s (Δt_w).

The switching from weak binding to strong binding or detachment can be expressed by the transition rates $k_{w \to s}$ and $k_{w \to d}$. The relative fractions, 86% and 14%, are proportional to these rates. Thus, based on our results, $k_{w \to s}$ was calculated as $(k_{w \to d} + k_{w \to s}, 1/310\,\mu\text{s}) \times 0.86 = 2.800\,\text{s}^{-1}$, which is 30-fold larger than that of weakly bound myosin suspended in solution and strain-free ($89\,\text{s}^{-1}$).[18] Consequently, the transition of weak binding to strong binding is greatly accelerated by rapid backward strain (see Table 2).

The weak binding to strong binding transition is thought to be coupled with Pi release from the myosin head.[18] If so, our results suggest that Pi release is accelerated by backward strain. Another scenario is worth considering, however. Markus et al. explored the sensitivity of the myosin head's structural change to force applied on the lever-arm by computer simulations.[19] Backward load induced strong binding by changing the conformation of the actin binding site in the myosin head (closure of actin binding cleft). Opening and closing of the cleft should be mechanically linked with the opening and closing of the exit route of Pi, termed the "backdoor".[20] Therefore, backward load applied on the lever-arm may directly induce strong binding by closure of this cleft, then, Pi release is accelerated or spontaneously released from myosin head.

Table 2.   Summary of mechano-sensitive weak binding for myosin-VI.

	Backward strain	Forward strain	Without strain
Short-lived weak binding: $k_{w \to d} + k_{w \to s}$	$3200\,\text{s}^{-1}$ ($310\,\mu$s)	$520\ \text{s}^{-1}$ (1.9 ms)	Not Det.
Weak to detach transition: $k_{w \to d}$	$448\,\text{s}^{-1}$ (2.2 ms)	$520\,\text{s}^{-1}$ (1.9 ms)	Not Det.
Weak to strong transition: $k_{w \to s}$	$2800\,\text{s}^{-1}$ ($357\,\mu$s)	N.D.	$30 - 89\,\text{s}^{-1}$ ($11 - 33$ ms)

4.2. *Strain sensor as a rectifier of Brownian motion*

We propose a model to explain how strong binding is accelerated in a strain-dependent manner (Fig. 5B). The opening and closing of the back door has been mechanically linked to that of the nucleotide (ATP or ADP)-binding pocket, or "front door".[21] The strain dependencies of the ADP release and ATP binding rates have previously been measured, revealing that external force applied to the head strains the front door. When the lever-arm is pulled backward, the head is bent forward, which closes the front door and opens the back door. Thus, backward strain accelerates Pi release and hence strong binding. At the same time, closing of the front door suppresses ADP release or ATP binding,[14] which means the overall ATP turnover rate is slow under backward load. On the other hand, upon forward strain, the backdoor is closed or unaffected, resulting in a relatively rare transition to strong binding. This strain-dependent asymmetric catch mechanism should be important for the Brownian head's reliable forward movement. We have thus named this mechanism the "strain-sensor mechanism".

To explicitly examine the relationship between the direction of myosin movement and the strain-sensor mechanism, the same scanning experiment was done for myosin-V, the reverse directional motor compared with myosin-VI. The results of the myosin-V experiments are in agreement with our model, which assumes backward strain accelerates strong binding (Fig. 5A). Therefore, the strain-sensor mechanism should contribute to the directionality of the myosin motor.

The strain sensor mechanism is notable in that random Brownian motion is rectified in one direction by sensing the intensity and direction of the mechanical strain. Using this simple mechanism, myosin can autonomously sense positional information and adaptably respond to changes in its environment. These characteristics should be important in motor assembly systems like muscle.

4.3. *Inhibition of ATP synthesis*

Mechano-chemical coupling for many linear and rotary motors has been studied at the single molecule level, which has shown that tight coupling is common for forward motion.[10,22] In contrast, the modes of force-induced backward motion seem to be quite diverse. Whereas for the rotary motor, F1-ATPase, the hydrolysis cycle is completely reversible and forced backward rotation can lead to ATP synthesis,[23] forced backward steps of myosin-V cannot synthesize ATP and are approximately irreversible.[24,25]

One possible reason for the irreversibility is the strain-sensor. Based on our model, because backward strain accelerates Pi release and always biases the hydrolysis cycle toward the ADP state to inhibit ATP synthesis, even in the presence of 50 mM Pi.[25] This irreversibility is physiologically reasonable because myosin-V and -VI function as a vesicle transporter in a cell, which contains milli-molar concentrations of ADP and Pi. Synthesizing ATP and moving backwards would severely compromise the efficiency of the myosin transport function.

5. Energetics of Myosin Motor

Above we describe two elementary mechanical processes in myosin function: the lever-arm swing and Brownian search-and-catch by a strain sensor mechanism. The two processes should contribute not only to the directionality of movement, but also the force generation because myosin-V and -VI can produce up to 2–3 pN against an external load. In conventional textbooks,[26] myosin generates force by a lever-arm swing (structural change). However, no reports have measured how large a force the lever-arm swing actually generates because conventional single molecule measurement assays cannot deconstruct the lever-arm swing from the Brownian search-and-catch. Recent technologies have overcome this limitation, revealing the contributions of the two mechanical processes and the total work done by a myosin motor.

5.1. Single molecule force measurement using DNA handle

Our team applied a DNA handle in our optical tweezers assay system.[27] The DNA handle is composed of double-stranded DNA ∼60 nm in length and tethers the myosin head directly to a 200-nm fluorescent polystyrene bead trapped by the system (Fig. 6A). The bead position was detected with nanometric and millisecond spatio-temporal resolution. Myosin-V was observed walking along actin filaments using this setup.[28]

In the presence of ATP, 77 nm processive steps that could be divided into 20 nm and 57 nm sub-steps were observed. Because the duration of the sub-steps was 123 ms, which is comparable to the first-passage time for a Brownian head to reach a forward actin binding site 77 nm away under load, and the expected bead displacement from the lever-arm swing is 20 nm,[29] we concluded the sub-steps to represent a 20 nm lever-arm swing and 57 nm Brownian search-and-catch.

These sizes would appear to disagree with the description above for 40-nm gold nanoparticles experiments, where, 51 nm and 25 nm sub-steps with a Brownian duration of 16 ms had been observed (Fig. 2). Differences in the experimental design can explain this discrepancy, as the drag coefficient for the optically trapped bead is 5-fold larger than the gold particle due to the difference in diameter, while the load in the DNA-handle setup pulled

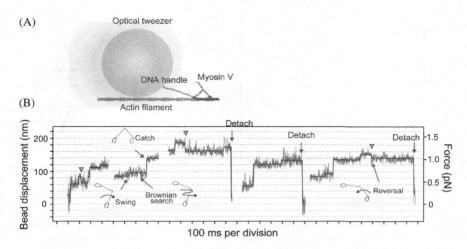

Fig. 6. Optical tweezers assay using a DNA handle. (A) Experimental setup. (B) Typical trajectories showing lever arm swing and the reversal.

the detached Brownian head backward and slowed its first-passage time exponentially.

5.2. *Fluctuation between lever-arm swing and the reversal under load*

We also observed $-20\,\text{nm}$ sub-steps that coupled with a $20\,\text{nm}$ lever-arm swing (Fig. 6B), suggesting these sub-steps are the result of a lever-arm reversal (Fig. 3, state $3 \rightarrow$ state 2). The duration of state 3 and state 2 in Fig. 3, τ_1 and τ_2, respectively, were load-dependent and fit to a single exponential curve. The inverse of the duration describes the transition rates

$$\frac{1}{\tau_1} = k_{\text{total 1}} = k_{\text{reversal}} + k_{\text{catch}} + k_{\text{detach 1}} \tag{1}$$

$$\frac{1}{\tau_2} = k_{\text{total 2}} = k_{\text{swing}} + k_{\text{detach 2}}. \tag{2}$$

Rearranging these equations and taking the ratio gives the frequency of the observed events ($k_{\text{reversal}} : k_{\text{catch}} : k_{\text{detach 1}} = N_{\text{reversal}} : N_{\text{catch}} : N_{\text{detach 1}}$ and $k_{\text{swing}} : k_{\text{detach 2}} = N_{\text{swing}} : N_{\text{detach 2}}$), where N_i is the number of observations for state i.

Figure 7A shows the load dependent transition rate for the lever-arm swing and lever-arm reversal. The lever-arm swing involves a structural change that is opposed by the load. Therefore, the reaction rate decreases

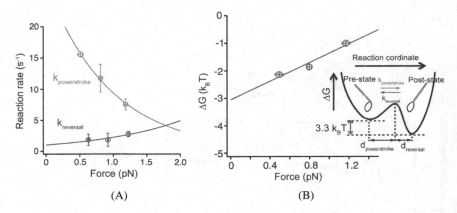

Fig. 7. Structural stability of lever arm for myosin-V. (A) Load dependent reaction rate for lever arm swing ($k_{\text{powerstroke}}$) and the reversal (k_{reversal}). (B) Load dependent free energy difference (ΔG) between the pre- and post-powerstroke states.

with increasing load in a manner that obeys an Arrhenius-type transition,

$$k(F) = k_0 e^{\left(\frac{Fd}{k_B T}\right)} \tag{3}$$

where k_0 is the transition rate in the absence of load, F is force, d is the characteristic distance, and $k_B T$ is the thermal energy.

For similar reasons, load should promote lever-arm reversal, which is consistent with our observation that the transition rate slightly increases with increasing load. The fitting parameter, d, which relates to the force dependency of the reaction, indicates the lever-arm reversal is less sensitive to load than the lever-arm swing. Therefore, once the lever-arm has swung, it robustly maintains the post-lever-arm swing conformation when sensing load.

5.3. Lever-arm swing versus Brownian search-and-catch

Given the individual rate constants, we can estimate the free-energy difference (ΔG) between the pre- and post-lever-arm swing states of the lead head using the formula

$$\frac{k_{\text{reversal}}}{k_{\text{swing}}} = e^{\left(\frac{\Delta G - Fd}{k_B T}\right)} \tag{4}$$

where F, Δx, k_B and T denote force exerted by the optical tweezers, the characteristic distance, Boltzmann's constant and absolute temperature, respectively. At no load, which describes the maximum energy bias of the lever-arm swing, ΔG was estimated to be $3.3\,k_B T$ (Fig. 7B).

The work done by a myosin dimer can be divided into two parts, work done by structural changes in the lever-arm and work done by the Brownian search-and-catch. We showed hand-over-hand stepping is triggered by a $3.3\,k_BT$ lever-arm swing in the lead head. The Brownian search-and-forward-catch contributes work to this system as load is exerted onto the myosin molecule. During physiological vesicle transport, myosin-V apparently distributes the load from a cargo bound to the tail domain completely onto the lead head. However, when the rear head undergoes Brownian motion, the lever-arm in the lead head should bend 10–15 nm backwards when a 2–3 pN load is exerted (bending stiffness $= \sim 0.\,2\,\text{pN/nm}$). In these cases, the Brownian head will conduct work when it catches an actin target 76 nm forward. Therefore, estimation of how much work the Brownian search-and-catch can potentially generate has physiological meaning. We found that the majority of myosin-V molecules completed the Brownian search-and-catch against 0.93 pN on average. This equates to a Brownian search-and-catch contribution of $13\,k_BT$ (57 nm $\times$ 0.9 pN) of work, which is similar to the chemical free energy released from the Pi release ($12\,k_BT$) when a myosin head strongly catches forward actin. Consequently, myosin-V predominantly works during the Brownian component by controlling Pi release using the strain-sensor mechanism, while the structural change of the lever-arm is just a trigger for the forward catch.

6. Physiological Advantages of the Brownian Machine

Figure 8 shows the contribution of the lever-arm swing and Brownian search-and-forward catch for total myosin-V work at various loads when assuming physiological vesicle transport geometry (vesicle is attached at tail domain). When 0.5 pN force is applied to the tail end, the lever-arm swing conducts $2.4\,k_BT$ (20 nm $\times$ 0.5 pN) of work, whereas the total work done by the myosin-V is $4.4\,k_BT$ (36 nm $\times$ 0.5 pN), meaning the Brownian search-and-catch component contributes $2.0\,k_BT$. Although the amount of work done by the lever-arm swing increases with load, it never exceeds $3.3\,k_BT$ at saturating physiological ATP concentration. Any residual work at high loads then is done by the Brownian search-and-catch. Thus, our results argue the myosin-V is a lever-arm-driven motor under low loads, but a Brownian search-and-catch driven motor under high loads. This result suggests that myosin-V can change its force-generating mechanism depending on the external force.

Knowing that the vast majority of work done by myosin-V at high load is a stochastic process (Brownian search-and-catch) and that the proportion of work done by the deterministic lever-arm swing and Brownian

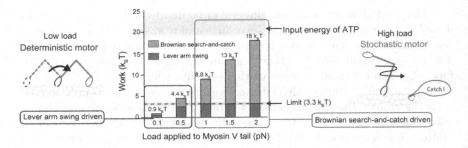

Fig. 8.   Mechanism for force generation at different loads.

search-and-catch varies with load offers important insights into the mechanisms used by myosin-V for its function. The hand-over-hand steps caused by deterministic lever-arm swings would be advantageous for smooth rapid movement at near-zero load when no obstacles are present. However, cells contain a cytoskeleton meshwork and a number of dynamic molecules and vesicles that risk disturbing transport. In these cases, the Brownian search-and-catch mechanism should be advantageous for avoiding such obstructions. Thus, our findings indicate that myosin-V is an optimized nanomachine highly adaptable to its intracellular environment, which should have significant implications on the design of artificial nanomotors.

Acknowledgments

I thank Keisuke Fujita for collaboration and Peter Karagiannis for reading the manuscript. This research has been supported by the Grant-in-Aid for Scientific Research on Innovative Areas of the Ministry of Education, Culture, Sports, Science and Technology (MEXT).

References

1. M. A. Hartman and J. A. Spudich, *J. Cell Sci.* **125**, 1627–1632 (2012).
2. F. Buss, G. Spudich and J. Kendrick-Jones, *Annu. Rev. Cell Dev. Biol.* **20**, 649-676 (2004).
3. Y. Y. Toyoshima, S. J. Kron, E. M. McNally, K. R. Niebling, C. Toyoshima and J. A. Spudich, *Nature* **328**, 536–539 (1987).
4. W. J. Greenleaf, M. T. Woodside and S. M. Block, *Annu. Rev. Biophys. Biomol, Struct.* **36**, 171–190 (2007).
5. T. Funatsu, Y. Harada, M. Tokunaga, K. Saito and T. Yanagida, *Nature* **374**, 555–559 (1995).

6. S. Nishikawa, I. Arimoto, K. Ikezaki, M. Sugawa, H. Ueno, T. Komori, A. H. Iwane and T. Yanagida, *Cell* **142**, 879–888 (2010).
7. H. Ueno, S. Nishikawa, R. Iino, K. V. Tabata, S. Sakakihara, T. Yanagida and H. Noji, *Biophys. J.* **98**, 2014–2023 (2010).
8. K. Shiroguchi and K. Kinosita Jr., *Science* **316**, 1208–1212 (2007).
9. E. M. De La Cruz, A. L. Wells, S. S. Rosenfeld, E. M. Ostap and H. L. Sweeney, *Proc. Natl. Acad. Sci. USA* **96**, 13726–13731 (1999).
10. M. Rief, R. S. Rock, A. D. Metha, M. S. Mooseker, R. E. Cheney and J. A. Spudich, *Proc. Natl. Acad. Sci. USA* **97**, 9482–9486 (2000).
11. S. Uemura and S. Ishiwata, *Nat. Struct. Biol.* **10**, 308–311 (2003).
12. C. Veigel, S. Schmitz, F. Wang and J. R. Sellers, *Nat. Cell Biol.* **7**, 861–869 (2005).
13. Y. Oguchi, S. V. Mikhailenko, T. Ohki, A. O. Olivares, E. M. De La Cruz and S. Ishiwata, *Proc. Natl. Acad. Sci. USA* **105**, 7714–7719 (2008).
14. D. Altman, H. L. Sweeney and J. A. Spudich, *Cell*, **116**, 737–749 (2004).
15. B. Brenner, *Proc. Natl. Acad. Sci. USA* **88**, 10490–10494 (1991).
16. M. Iwaki, A. H. Iwane, T. Shimokawa, R. Cooke and T. Yanagida, *Nat. Chem. Biol.* **5**, 403–405 (2009).
17. M. Iwaki, H. Tanaka, A. H. Iwane, E. Katayama and T. Yanagida, *Biophys. J.* **90**, 3643–3652 (2006).
18. E. M. De La Cruz, E. M. Ostap and H. L. Sweeney, *J. Biol. Chem.* **276**, 32373–32381 (2001).
19. M. Duttmann, Y. Togashi, T. Yanagida and A. S. Mikhailov, *Biophys. J.* **102**, 542–551 (2011).
20. R. G. Yount, D. Lawson and I. Rayment, *Biophys. J.* **68**, 44S–47S (1995).
21. K. C. Holmes, I. Angert, F. J. Kull, W. Jahn and R. R. Schroder, *Nature* **425**, 423–427 (2003).
22. B. E. Clancy, W. M. Behnke-Parks, J. O. Andreasson, S. S. Rosenfeld and S. M. Block, *Nat. Struct. Mol. Biol.* **18**, 1020–1027 (2011).
23. Y. Rondelez, G. Tresset, T. Nakashima, Y. Kato-Yamada, H. Fujita, S. Takeuchi and H. Noji, *Nature* **433**, 773–777 (2005).
24. J. C. Gebhardt, A. E. Clemen, J. Jaud and M. Rief, *Proc. Natl. Acad. Sci. USA* **103**, 8680–8685 (2006).
25. A. R. Dunn and J. A. Spudich, *Nat. Struct. Mol. Biol.* **14**, 246–248 (2007).
26. B. Alberts, A. Johnson, J. Lewis, M. Raff, K. Roberts and P. Walter, *Molecular Biology of the Cell*, 5th edn. (Garland Science, USA, 2008).
27. N. R. Guydosh and S. M. Block, *Nature* **461**, 125–128 (2009).
28. K. Fujita, M. Iwaki, A. H. Iwane, L. Marcucci and T. Yanagida, *Nat. Commun.* **3**(956), 1–9 (2012).
29. P. D. Coureux, A. L. Wells, J. Menetrey, C. M. Yengo, C. A. Morris, H. L. Sweeney and A. Houdusse, *Nature* **425**, 419–42 (2003).

Chapter 5

DIFFUSIOPHORETIC NANO AND MICROSCALE PROPULSION AND COMMUNICATION

Vinita Yadav, Wentao Duan and Ayusman Sen*

Department of Chemistry
The Pennsylvania State University
University Park, Pennsylvania 16802, USA
* asen@psu.edu

1. Introduction

Nano and microscale propulsion is ubiquitous in nature.[50-53] Unicellular organisms like bacteria are not only motile but can also sense food/toxin gradients, then interact/communicate amongst themselves and respond accordingly.[51] Thus, the three basic requirements to build dynamic and responsive nano and microscale machines and robots are (1) motility driven by local energy harvesting, (2) ability to sense each other and the environment, and (3) ability to respond collectively. As with living systems, a continuous input of energy and information is necessary. We aim to address these three issues in this chapter. As we go down the length scale, the laws of physics begin to change.[54] As the radius of an object scales down, the decrease in volume is exponentially higher than the decrease in surface area. This implies volume dependent forces like inertia, which dominate higher up the scale, lose relevance as we scale down. Instead, it is the surface forces that need to be channeled in order to induce motion.[13]

There are certain other factors that also need to be considered in order to engineer micro- and nanoscale motion and these are discussed below.

1.1. *Reynold's number and Brownian motion*

Reynold's number (Re) refers to a dimensionless quantity often invoked when performing scaling of fluid dynamics problems. It is defined as the ratio of the inertial and viscous forces and helps to predict/characterize fluid patterns under different fluid conditions:

$$Re = \rho V l / \eta \tag{1}$$

where ρ is the density of the fluid, V is the mean velocity relative to the fluid, l is the characteristic linear dimension or the travelled length of the fluid and η is the dynamic viscosity of the fluid. Laminar flows occur at low Reynolds numbers, where viscous forces are dominant, and are characterized by smooth, constant fluid motion while turbulent flows occur at high Reynolds numbers and are dominated by inertial forces. This principle is often used in designing microfluidic devices. In our case, it helps to determine which of the two forces — inertial or viscous dominates. Bacteria and other unicellular organisms are the finest examples of low Reynolds number swimmers (Re $= 10^{-4}$). For comparison, an average sized human being has a Reynolds's number of 10^4. Inducing motion at low Reynolds number also requires introducing asymmetry in the object to evade reciprocal motion, in accordance with the scallop theorem.[55]

Low Reynolds number represents the first challenge to nano- and microscale motion. However, it is not the only one. We know from classical statistical mechanics that every molecule moves randomly in all three dimensions with average kinetic energy of $KT/2$; K being the Boltzmann constant and T the absolute temperature. This random movement causes collision between different molecules and gives rise to Brownian motion generating both translational and rotational diffusion. Translational particle diffusion caused due to Brownian motion can be calculated as:

$$D_t = kT/6\Pi\eta a \tag{2}$$

while the rotational particle diffusion is given by:

$$D_r = kT/(8\Pi\eta a^3) \tag{3}$$

where D_t is the translational diffusion coefficient and D_r is the rotational diffusion coefficient of the particle, k is the Boltzmann constant, T is the absolute temperature, η is the viscosity of fluid through which the particle moves and a is the radius of such a particle.

Inducing directed motion at the nano- and micron-scale requires overcoming these randomizing events and various mechanisms have been examined to accomplish this. The next section focuses on one such mechanism.

2. Mechanisms of Motility

One of the first mechanisms identified for autonomous motion was self-electrophoresis. In this context, Anderson recognized a critical concept, the slip velocity, at the solid–liquid interface and the role it plays in fluid dynamics, thus, laying down the foundation for such phoretic transport mechanism.[48,49] At low Reynold's number, where surface forces dominate, it is often processes occurring within this thin interfacial layer that control the fluid dynamics. In a solution, the charge on a particle's surface is balanced by a diffuse cloud of counter ions (Fig. 1). The charge density $\rho_e(y)$, within the cloud at a distance y, decays exponentially in y at distances of order of the Debye screening length K^{-1} from the surface. Taken together, the surface charge and the diffuse cloud, called the "double layer," are a neutral body. The Debye length plays an important role in controlling the behavior of colloidal particles and is defined as:

$$K^2 = (2Z^2e^2c_\infty)/\varepsilon kT \tag{4}$$

where Z is the absolute value of the valency of the ion, e is the charge on an electron, and c is the concentration of the ions ε is defined as the dielectric constant of the material, k is the Boltzmann constant, T is the absolute temperature.

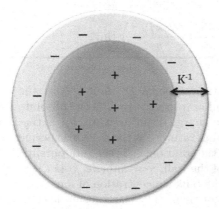

Fig. 1.   The electric double layer of a charged particle in a polar solution. The counter ions from the solution come near the charged particle surface to neutralize the charge and this fluid layer remains diffused around the particle. K^{-1}, the thickness of the double layer, is defined as the Debye screening length and is dependent on the concentration of ions in the surrounding fluid. A non-spherical charged surface behaves the same way.

While low ionic strengths lead to high Debye lengths resulting in colloidal stability, a high ionic strength solution implies a small Debye length, which leads to short range van der Waals forces dominating and leads to particle aggregation.

Phoretic transport is defined as the movement of colloidal particles by a field that interacts with the surface of each particle;[48,49,56,57] for instance, electrophoresis involves an electric field gradient, thermophoresis involves a thermal gradient[58-60] and diffusiophoresis involves a gradient of ionic or non-ionic chemical species.[27,61,62] Other mechanisms like propulsion based on Marangoni effect,[63-66] bubble propulsion,[67-70] as well as propulsion under magnetic[71-78] or acoustic fields[79-81] have also been identified.

2.1. *Electrolyte diffusiophoresis*

Electrolyte diffusiophoresis operates when a gradient of electrolytes is formed across a charged surface. For diffusiophoresis near a wall, there are two effects contributing to the movement of a particle: an electrophoretic effect and a chemophoretic effect, and the speed of the diffusiophoretic movement can be approximated by the equation below:[82-84]

$$
\mathbf{U} = \underbrace{\frac{\nabla c}{c_0} \left[\left(\frac{D^+ - D^-}{D^+ + D^-} \right) \left(\frac{k_B T}{e} \right) \frac{\varepsilon(\zeta_p - \zeta_w)}{\eta} \right]}_{\text{Electrophoretic Term}}
$$
$$
+ \underbrace{\frac{\nabla c}{c_0} \left[\left(\frac{2\varepsilon k_B^2 T^2}{\eta e^2} \right) \{ \ln(1 - \gamma_w^2) - \ln(1 - \gamma_p^2) \} \right]}_{\text{Chemophoretic Term}}
$$

(5)

where U is the particle velocity, D^+ and D^- are the diffusion coefficients of the cation and anion respectively, Z is the absolute value of the valences of the ions, e is the charge of an electron, k_B is the Boltzmann constant, T is the absolute temperature, ε is the dielectric permittivity of the solution, η is the viscosity of the solution, ζ_p is the zeta potential of the particle, ζ_w is the zeta potential of the wall, $\gamma = \tanh(Ze\zeta_p/4kT)$, ∇c is the concentration gradient and c_0 is the bulk concentration of ions at the particle location, as if the particle was not there. The electroosmotic component, caused due to the wall double layer, is given by a similar equation, with the particle zeta potential replaced by the wall zeta potential.

The two parts of the equation signify the two components/contributions to diffusiophoresis, as shown in Fig. 2. The first half signifies electrophoresis. The electric field in this case originates from the difference in diffusion between the cation and anion which contributes to the ion gradient in a given direction. This difference leads to a net electric field, which acts both

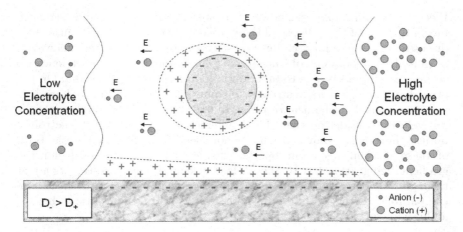

Fig. 2. Schematic depiction of diffusiophoretic motion. The difference in diffusivity of the ions generated from the source causes a local electric field. The double layer around the particles as well as the wall responds to the thus formed electric field leading to electrophoretic and electroosmotic motion respectively. In the example in the figure above, the anion diffuses faster than the cation generating an electric field from right to left. The electrophoretic motion of a negatively charged particle is from left to right. Correspondingly, the electroosmotic flow along the negatively charged wall is from right to left. The concentration gradient also leads to thickness gradient of double layers on the surfaces of the particle and wall, and in-turn a pressure difference that propels particles from left to right.

electrophoretically on the nearby particles and electroosmotically on the ions adsorbed in the double layer of the wall. The electroosmotic component leads to fluid movement near the walls. Depending on the charge of the particle, the electrophoretic and electroosmotic components can augment or allay each other. In the case of competition between the two, the zeta potential of particle or wall dominates and leads to reduced velocities. However, when both electroosmotic and electrophoretic motion are in the same direction, an enhancement in particle speed is observed. Interplay between the osmotic and phoretic components can also lead to schooling and exclusion patterns as discussed in Sec. 3.2.1.

The second component is the chemophoretic effect. The concentration gradient of the electrolytes causes a gradient in the thickness of the electric double layer, and thus a "pressure" difference along the wall is created. As a result, the solution will flow from the area of higher electrolyte concentration to that of lower concentration, known as the chemophoretic effect.

The combination of electrophoretic and chemophoretic effects leads to an overall diffusiophoretic flow, which powers the movement of particles.

Electrolyte diffusiophoresis, however, is not effective in high ionic strength media because of the collapse of the double layer on the particle surface, as discussed in the previous section. Non-electrolyte diffusiophoresis, however, is caused by a gradient of uncharged solutes and has no dependence on surface charge and is able to function in high ionic strength media. On the other hand, in a low ionic strength medium, electrolyte diffusiophoresis is a more powerful mechanism resulting in higher speeds. This is shown qualitatively by considering that the chemophoretic component of electrolyte diffusiophoresis has similar origins as non-electrolyte diffusiophoresis. Both of these mechanisms occur by the chemical species responsible for the gradient being attracted to the surface either by electrostatic (ionic) or through van der Waals (non-ionic) interactions. If these two effects are comparable, the electrolyte diffusiophoresis is stronger because it has an additional electric field term (Eq. (5)).

3. Diffusiophoresis-Based Systems

Diffusiophoresis-based systems have been utilized to study collective behavior between particles. Micro/nano-sized particles which produce attractive/repulsive diffusiophoretic interactions between them, allowing them to show corresponding "schooling"/"exclusion" patterns, have been extensively studied.[83,86−88] Immobilized "motors" can transfer their force to the surrounding fluid; in effect, functioning as micro-pumps. Thus, the diffusiophoretic mechanism can be employed to design triggered, self-propelled micro-pumps.[84,85]

The trigger/gradient in a diffusiophoretic system can be either externally or self-generated. The latter refers to motion in response to a gradient generated by the motile particles themselves. The following sections expand on these two facets, providing examples.

3.1. *Externally triggered diffusiophoretic systems*

3.1.1. *"On/off" micro-pump and photo-colloidal diode*

An external control switch is important for it allows the system to respond to changes in the environment, which is useful for the design of sensors and logic gates. Previously reported motors/pumps typically lack control, i.e. they function till all the fuel is consumed. As described below, it is now possible to design micro-pumps that are capable of being turned on/off by a specific external signal.[84]

Light was employed to trigger ion formation from a photoacid. As depicted in Fig. 3, upon illumination of a solid photoacid generator,

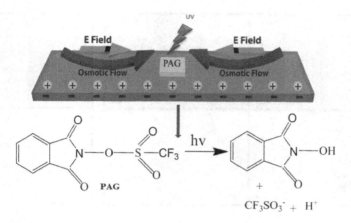

Fig. 3. Schematic depiction of PAG pumping mechanism. The negative surface charge of the glass creates a positive double layer, which in response to the generated ions causes an inward electroosmotic flow. The negatively charged tracers (S-PS particles) move opposite to the direction of the electric field, competing against the electroosmotic flow while the positively charged tracers (NH$_2$-PS particles) move along the electric field direction aided by the electroosmotic flow.

N-hydroxyphthalimide triflate (PAG), at wavelength 365 nm, it decomposes to form N-hydroxyphthalimide, and two ions: proton and triflate anion. With the proton diffusing faster ($D = 9.31 \times 10^{-5}\,\mathrm{cm^2\,s^{-1}}$) than the larger triflate anion (estimated $D = 1.38 \times 10^{-5}\,\mathrm{cm^2\,s^{-1}}$ assuming a sphere), a local electric field is set up pointing inwards. Owing to the electric double layer on the negatively-charged sodium borosilicate glass slide, an electroosmotic flow is generated, which is also inwards in the direction of the local electric field. Positively charged tracers (amino functionalized polystyrene particles, NH$_2$-PS) move towards the photoacid, aided by both the diffusiophoretic and electroosmotic flows while the negatively charged tracers (sulfate functionalized polystyrene particles; S-PS) move in the direction of the diffusiophoretic flows, out-winning the electroosmotic flow, owing to their higher zeta potential in comparison to that of the glass.

The motion ceases when UV light is turned off but is re-initiated upon re-illumination. Figure 4 displays the tracer particle distributions without UV and after 1 min of UV irradiation. This micro-pump also enables self-assembled patterns as shown in Figs. 4E and 4F.

In order to separate the diffusiophoretic and electroosmotic components, the experiments were repeated on a polystyrene surface, which has minimal surface charge. The velocities of the positively charged tracers were impeded in the absence of the aiding electroosmotic force, while those of the negatively charged tracers were enhanced due to the absence of the opposing

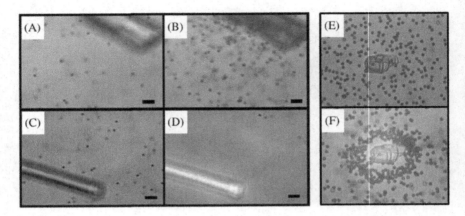

Fig. 4. Optical microscope images of particle motion. (A) and (B) show the distribution of the positively charged tracers (NH$_2$-PS) around the photoacid (PAG) microcrystallites with UV off (control) and after 1 min of UV illumination respectively. (C) and (D) display the same for the negatively charged tracers (S-PS). Scale bar is 10 μm. (E) and (F) display control (E) and the self-assembled pattern (F) formed due to the diffusiophoretic motion.

electroosmotic force, both approximately by a factor of two. Thus, the estimated contributions of diffusiophoretic and electroosmotic components to velocity were considered equal.

Another micro-pump was designed based on externally triggered acid-catalyzed hydrolysis of a polymeric imine, poly(4-formyphenyl acrylate) aniline Schiff base (PFA-S). Only this time, the direction of the local electric field was reversed due to higher diffusivity of the anion ($D_{\text{Cl}^-} = 2.032 \times 10^{-5} \, \text{cm}^2 \, \text{s}^{-1}$) relative to the much larger cation (Fig. 5). Accordingly, the movement of the tracers was observed to be reversed with the negative S-PS tracers moving inwards competing against the electroosmotic component and the positively charged NH$_2$-PS particles moving outwards, aided by the electroosmotic component.

In order to induce control over fluid pumping, varying concentrations of HCl were introduced into the PFA-S system and in accordance with Eq. (5), the velocities attained by the tracer particles increased with increasing concentration gradient of the formed electrolytes. The particles showed maximum velocity at the lowest experimental pH; 1 M HCl as shown in Fig. 5.

Similar to a source-drain set up, the two externally triggered micro-pumps, generating opposite electric-field were combined to create a colloidal photo-diode which uses UV as the input to regulate the direction and speed of particle transport. Upon UV illumination, the PAG initiates the diffusiophoretic motion pushing the negatively charged tracers away in all

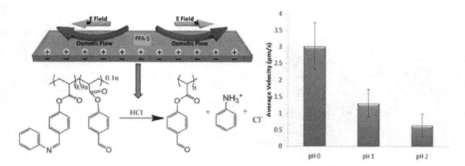

Fig. 5.   Schematic depiction of PFA-S pumping mechanism. The local electric field points outwards away from the polymer film and the negatively charged tracers S-PS particles move inwards, towards film. The right panel shows the velocity distribution histograms of negatively charged tracers as a function of the acid concentration for the PFA-S pump. The ionic strength at each acid concentration (1 M, 0.1 M, 0.01 M HCl) was kept the same at 1 M, to avoid any changes to the electric double layer.

directions. When the acid formed by the photolysis of PAG reaches the PFA-S film, it starts imine hydrolysis; the PFA-S film then actively pulls the tracers precisely towards itself, and further enhances the tracer velocities. This push–pull mechanism results in rectification and amplification of particle motion.

Figure 6 depicts the spatial and temporal control achieved using the source-drain set-up, with the particles reaching a maximum average velocity of 3.9 μm/s. Effectively, the colloidal photo-diode is capable of amplifying the velocity, as well as directing the particle flow in one direction, without introducing a third "base/gate" terminal as required by a typical transistor. This opens the door to creating more complex colloidal logic systems.

3.1.2. *Triggered crack-detection, targeting and repair using ion gradients*

In another example of an externally triggered diffusiophoretic system, drug loaded nanoparticles were delivered to repair damaged sites. The movement of the repair/healing agents was triggered by an ion gradient created by the damaged entity. Thus, ions emanating from the damaged site act both as the trigger and the fuel.[47] Demineralization due to a crack or damage in bone results in ion leaching; specifically, hydroxyapatite demineralizes into calcium, dihydrogen phosphate and hydroxide ions and the high diffusivity of hydroxide ion (OH^-) induces a local electric field pointing away from the crack, inducing anionic particles to move towards it (Fig. 7). This technique,

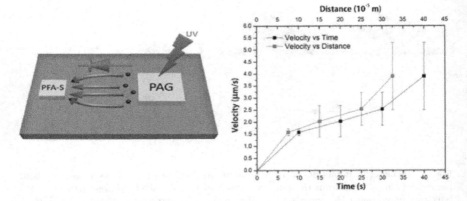

Fig. 6. Schematic depiction of source (PAG)-drain(PFA-S) based colloidal photo-diode indicating both rectification and amplification of the movement of tracer particles. Right panel displays spatial and temporal regulation of velocity (S-PS particles) attained using the source-drain photo-diode. Distance is measured from edge of the PAG and time is measured from when the UV is turned on.

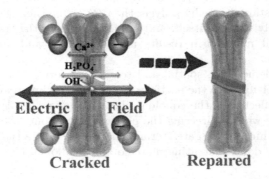

Fig. 7. Schematic depiction of the electric field generated due to the leaching ion-gradient from the cracked bone and the resulting migration of anionic particles. The particles loaded with drug actively target the crack site and promote healing.

for the first time enables the design of suitably charged therapeutic vehicles that can actively migrate to the damage site to aid healing.

The delivery mechanism was first tested with charged quantum dots and anionic proteins. Sodium alendronate, a known osteoporotic drug, loaded particles were then synthesized that could be triggered by the electric field generated due to the ion gradient from the crack. The synthesized particles were also tested on human MG-63 cells — an immortalized osteoblast cell line using an *in-vitro* cell proliferation assay that confirmed increased cell growth (~10%).

3.2. Self-triggered diffusiophoretic system: Collective behaviors of micromotors in response to orthogonal stimuli

In contrast to motility in response to external gradients, particles can generate their own gradient and respond by moving. Furthermore, the overlapping gradients from a group of particles can result in collective behavior. In general, collective behavior relies on the fast and precise communication and interaction between individual agents. Examples in living systems include fish schooling and bird flocking. Inspired by their natural counterparts, several synthetic systems have been designed to exhibit collective behavior: In response to UV light,[82,86] silver chloride microparticles attract each other and cluster to form "schools". As expected, the motility of the particles changes depending on the number neighboring particles.[89] Gold microparticles in hydrogen peroxide solution were also found to swarm with addition of hydrazine.[90] These systems exhibit a single collective behavior in response to a single stimulus, the supply/drainage of fuel (chemical) or energy (light, electric or magnetic field). Recently, a silver phosphate (Ag_3PO_4)-based micromotor system has been described which exhibits two types of collective behavior, namely "schooling" and "exclusion", with the system reversibly transitioning between the two in response to the change of external stimulus. Based on the system's response to two orthogonal stimuli, a NOR logic gate was constructed with stimuli as inputs and collective behavior as outputs.

3.2.1. Reversible transition between "exclusion" and "schooling"

Silver phosphate particles that sit above a negatively-charged surface transition reversibly between "exclusion" and "schooling" behavior in response to addition/removal of ammonia, as shown in Fig. 8. With the addition of ammonia, reaction (6) shifts towards right. Since the OH^- ion diffuses away from the particle surface much faster than the other, larger, ions, the generated ion gradients lead to outward electrical field, as shown in Fig. 8 bottom left. As discussed previously, the interactions between particles are governed by the direction of electric fields as well as zeta potential difference between the particles and the wall. Outward electric fields lead to inward electrophoresis of negatively charged silver phosphate as well as outward electroosmosis along the surface of negatively charged glass. Because the magnitude of the Ag_3PO_4 particle zeta potential ($\zeta_p = -55\,\text{mV}$) is smaller

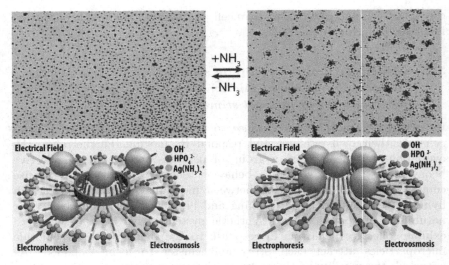

Fig. 8.   Scheme demonstrating mechanism of transition between two collective behaviors in response to ammonia addition/removal.

than that of the glass slide ($\zeta_w = -85\,\mathrm{mV}$),[91] electroosmosis dominates, particles repel each other and "exclusion" emerges in consequence.

When ammonia is removed from the system, reaction (6) shifts towards left. With the ions migrating inwards, direction of the electric field is reversed as well as that of the diffusiophoretic flows. With inward diffusiophoretic flows, particles move towards each other and cluster to form small "schools". These "schools" also attract each other, and merge into larger schools, as shown in Fig. 9.

$$Ag_3PO_4 + 6\,NH_3 + H_2O \rightleftharpoons 3\,Ag(NH_3)_2^+ + HPO_4^{2-} + OH^-. \qquad (6)$$

In addition to interaction between Ag_3PO_4 particles, interaction between active Ag_3PO_4 particles and inert negatively charged tracer particles was also investigated. Again, when ammonia was added to the system, equilibrium (6) shifts to the right, and the resulting ions secreted from the active Ag_3PO_4 particles generate diffusiophoretic flows that cause "exclusion" zones to form not only between active particles, but also between the active and inert ones (in this case, $0.9\,\mu\mathrm{m}$ PS-carboxylate tracer particles). As a result, the PS-carboxylate tracers are pumped away from large Ag_3PO_4 particles that sit on the glass slide. Also, probably due to the difference in reactivity, the size of exclusion zones formed increase with increasing size of the Ag_3PO_4 "pump" particles (Fig. 10). On the other hand, with ammonia removed from the system, the ion concentration gradient and the directions of diffusiophoretic flow are reversed. As a result,

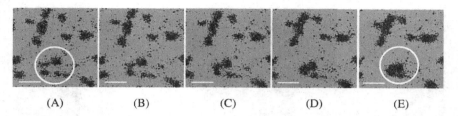

(A)   (B)   (C)   (D)   (E)

Fig. 9. Time-lapse optical microscope images demonstrating dynamic merging of smaller "schools" to form larger ones. (A)–(E) Four small schools within the white circle gradually merge with each other to form a large one: (A) 0 s, (B) 10 s, (C) 25 s, (D) 40 s, (E) 60 s. Scale bar, 20 μm.

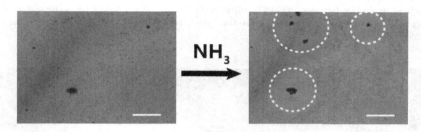

Fig. 10. Time-lapse optical microscope images showing silver orthophosphate microscale pump system with 0.9 μm PS-carboxylate tracer particles in 2 mM ammonia solution. With addition of ammonia solution, tracer particles are pumped away from the large pump particles and "exclusion" zones (areas within the white circles) are formed within seconds. The sizes of "exclusion" zones vary from pump particles. Scale bar, 50 μm.

the active silver phosphate particles attract each other as well as the silica tracers, and form "schools". It is interesting to note that the inert silica tracers mainly sit on the outer boundary of the formed "schools", and when the smaller "schools" start to fuse, the silica tracers that are at the merging boundaries are "excluded" to the periphery of the newly formed larger "school". Thus, with smaller schools merging altogether, hierarchical particle assembly of active Ag_3PO_4 particles and inert silica tracers is achieved (Fig. 11).

3.2.2. *"Exclusion" in response to UV light*

Silver phosphate particles also respond to UV light, with the following reaction taking place:

$$(4x + 4y)Ag_3PO_4 + (4y - 2x)H_2O$$
$$\leftrightharpoons 12x\ Ag + 12y\ Ag^+ + 3x\ O_2 + (4y - 8x)OH^- + (4x + 4y)HPO_4^{2-}. \quad (7)$$

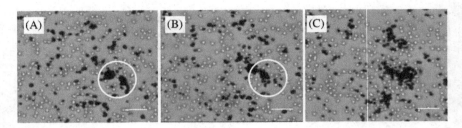

Fig. 11. Time-lapse optical microscope images showing hierarchical particle assembly between Ag_3PO_4 microparticles and 2.34 μm silica tracers in 2 mM ammonia solution. (A) With removal of ammonia, Ag_3PO_4 microparticles forms "schools" with silica tracers on the outer boundary; (B) when the smaller "schools" start to merge, silica tracers are "excluded" to the periphery of the newly formed larger "schools". (C) Hierarchical particle assembly between Ag_3PO_4 microparticles and silica tracers is achieved with Ag_3PO_4 microparticles "schools" as the core and silica tracers on the periphery. Scale bar, 20 μm.

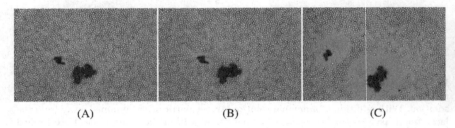

Fig. 12. Time-lapse optical microscope images demonstrating formation of exclusion zones around silver phosphate particles upon UV exposure (A) before UV exposure (B) UV turned on, (C) 15 s after UV exposure. Scale bar, 20 μm.

Upon UV exposure, ions are generated through reaction (7) and the faster diffusivity of OH^- leads to an outward electric field. Similar to the case described in Sec. 3.2.1, outward electric field leads to repulsion between silver phosphate particles and nearby silica tracers, and exclusion zones form around the silver phosphates, as shown in Fig. 12.

3.2.3. Design of logic gate based on orthogonal stimuli

Now that we have two stimuli for the Ag_3PO_4 microparticles/ammonia system, a NOR gate can be designed with UV and ammonia as inputs, while "schooling" and "exclusion" behaviors as output 1 and 0, respectively, as shown in Fig. 13. After the formation of "schools" in the Ag_3PO_4 microparticles/ammonia system, without inputs of UV or ammonia, the "schooling" behavior results with ammonia removal and "1" as the output,

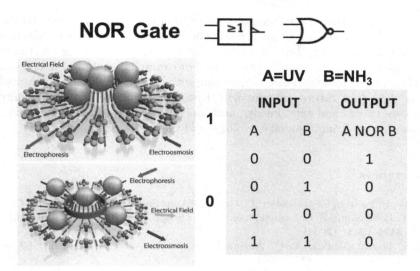

	A=UV	B=NH$_3$	
	INPUT		**OUTPUT**
	A	B	A NOR B
	0	0	1
	0	1	0
	1	0	0
	1	1	0

Fig. 13. Scheme demonstrating design of NOR Gate with UV and ammonia as inputs, collective behaviors as outputs: "schooling" and "exclusion" behaviors as 1 and 0, respectively.

while any input, either UV or ammonia, causes a shift to "exclusion" behavior and output of "0". The NOR gate design, as shown by Pierce, can be used to reproduce the functions of all the other logic gates,[92] and therefore perform any logical operation required.

4. Conclusion

The applications of diffusiophoretic nanoscale propulsion and interaction range from the design of novel active materials and systems to therapeutic agents. The grand challenge in this area is the rational design of populations of self-powered synthetic materials that have the ability to organize themselves, based on signals from each other and from their environment, to perform complex tasks. This will require an in-depth understanding of how energy and information processing can be designed into materials capable of exhibiting non-equilibrium, mesoscopic behaviors such as (a) dynamic hierarchical self-organization, (b) rapid reconfigurability and self-healing, and (c) collective and cooperative functions (e.g., targeted "on-demand" delivery of cargo). For the systems described in this chapter, the motility arises from the electric field derived from ion gradients that are either imposed externally or are self-generated. The ion gradients also serve as "information" that allows the particles to interact with each other and the environment, resulting in such emergent behaviors as directed collective motion and schooling. Much remains to be learned in integrating functional components into these active materials. For example, what are the optimal designs that combine information sensing with specific functional outcomes? It is clear that future progress will depend critically on a close synergy between theory and experiments, and expertise drawn from many different fields ranging from chemistry, biology, and physics, to engineering.

References

1. W. F. Paxton, K. C. Kistler, C. C. Olmeda, A. Sen, S. K. St. Angelo, Y. Cao, T. E. Mallouk, P. E. Lammert and V. H. Crespi, *J. Am. Chem. Soc.* **126**(41), 13424–13431 (2004).
2. S. Fournier-Bidoz, A. C. Arsenault, I. Manners and G. A. Ozin, *Chem. Commun.* **4**, 441–443 (2005).
3. G. A. Ozin, I. Manners, S. Fournier-Bidoz and A. Arsenault, *Adv. Mater.* **17**(24), 3011–3018 (2005).
4. S. Sánchez and M. Pumera, *Chem. Asian J.* **4**(9), 1402–1410 (2009).
5. A. Sen, M. Ibele, Y. Hong and D. Velegol, *Faraday Discuss.* **143**, 15–27 (2009).
6. J. Wang, *ACS Nano* **3**(1), 4–9 (2009).
7. S. J. Ebbens and J. R. Howse, *Soft Matter* **6**(4), 726–738 (2010).
8. T. Mirkovic, N. S. Zacharia, G. D. Scholes and G. A. Ozin, *Small* **6**(2), 159–167 (2010).
9. Y. Mei, A. A. Solovev, S. Sanchez and O. G. Schmidt, *Chem. Soc. Rev.* **40**(5), 2109–2119 (2011).
10. J. Wang and W. Gao, *ACS Nano* **6**(7), 5745–5751 (2012).
11. W. Duan, R. Pavlick and A. Sen, *Engineering of Chemical Complexity*, eds Mikhailov A. and Ertl G., Vol. 6, "Biology of nanobots," (World Scientific Publishing Company, Singapore), 125–144 (2013).
12. D. Patra, S. Sengupta, W. Duan, H. Zhang, R. Pavlick and A. Sen, *Nanoscale* **5**(4), 1273–1283 (2013).
13. W. Wang, W. Duan, S. Ahmed, T. E. Mallouk and A. Sen, *Nano Today* **8**(5), 531–554 (2013).
14. T. E. Mallouk and A. Sen, *Sci. Am.* **300**(5), 72–77 (2009).
15. S. Sengupta, M. E. Ibele and A. Sen, *Angew. Chem. Int. Ed.* **51**(34), 8434–8445 (2012).
16. R. Kapral, *J. Chem. Phys.* **138**, 020901 (2013).
17. T. Mirkovic, N. S. Zacharia, G. D. Scholes and G. A. Ozin, *ACS Nano* **4**(4), 1782–1789 (2010).
18. T. Sakaue, R. Kapral and A. S. Mikhailov, *Eur. Phys. J. B* **75**(3), 381–387 (2010).
19. G. M. Whitesides, *Sci. Am.* **285**(3), 78–83 (2001).

20. T. R. Kline, W. F. Paxton, T. E. Mallouk and A. Sen, *Angew. Chem. Int. Ed.* **44**(5), 744–746 (2005).
21. T. R. Kline, W. F. Paxton, Y. Wang, D. Velegol, T. E. Mallouk and A. Sen, *J. Am. Chem. Soc.* **127**(49), 17150–17151 (2005).
22. W. F. Paxton, P. T. Baker, T. R. Kline, Y. Wang, T. E. Mallouk and A. Sen, *J. Am. Chem. Soc.* **128**(46), 14881–14888 (2006).
23. Y. P. He, J. S. Wu and Y. P. Zhao, *Nano Lett.* **7**(5), 1369–1375 (2007).
24. J. R. Howse, R. A. L. Jones, A. J. Ryan, T. Gough, R. Vafabakhsh and R. Golestanian, *Phys. Rev. Lett.* **99**(4), 048102 (2007).
25. L. D. Qin, M. J. Banholzer, X. Y. Xu, L. Huang and C. A. Mirkin, *J. Am. Chem. Soc.* **129**(48), 14870–14871 (2007).
26. S. T. Chang, E. Beaumont, D. N. Petsev and O. D. Velev, *Lab Chip* **8**(1), 117–124 (2008).
27. U. M. Córdova-Figueroa and J. F. Brady, *Phys. Rev. Lett.* **100**(15), 158303 (2008).
28. D. Pantarotto, W. R. Browne and B. L. Feringa, *Chem. Commun.* **13**, 1533–1535 (2008).
29. J. G. Gibbs and Y. P. Zhao, *Small* **5**(20), 2304–2308 (2009).
30. P. H. Jones, F. Palmisano, F. Bonaccorso, P. G. Gucciardi, G. Calogero, A. C. Ferrari and O. M. Marago, *ACS Nano* **3**(10), 3077–3084 (2009).
31. Y. G. Tao and R. Kapral, *J. Chem. Phys.* **131**(2), 024113 (2009).
32. N. S. Zacharia, Z. S. Sadeq and G. A. Ozin, *Chem. Commun.* **39**, 5856–5858 (2009).
33. I.-K. Jun and H. Hess, *Adv. Mater.* **22**(43), 4823–4825 (2010).
34. G. Loget and A. Kuhn, *J. Am. Chem. Soc.* **132**(45), 15918–15919 (2010).
35. S. Sattayasamitsathit, W. Gao, P. Calvo-Marzal, K. M. Manesh and J. Wang, *Chemphyschem* **11**(13), 2802–2805 (2010).
36. A. A. Solovev, S. Sanchez, M. Pumera, Y. F. Mei and O. G. Schmidt, *Adv. Funct. Mater.* **20**(15), 2430–2435 (2010).
37. W. Gao, K. M. Manesh, J. Hua, S. Sattayasamitsathit and J. Wang, *Small* **7**(14), 2047–2051 (2011).
38. R. Liu and A. Sen, *J. Am. Chem. Soc.* **133**(50), 20064–20067 (2011).
39. G. Loget and A. Kuhn, *Nat. Commun.* **2**, 535 (2011).
40. J. L. Moran and J. D. Posner, *J. Fluid Mech.* **680**, 31–66 (2011).
41. O. S. Pak, W. Gao, J. Wang and E. Lauga, *Soft Matter* **7**(18), 8169–8181 (2011).
42. H. Zhang, K. Yeung, J. S. Robbins, R. A. Pavlick, M. Wu, R. Liu, A. Sen and S. T. Phillips, *Angew. Chem. Int. Ed.* **51**(10), 2400–2404 (2012).
43. J. Palacci, S. Sacanna, A. P. Steinberg, D. J. Pine and P. M. Chaikin, *Science* **339**(6122), 936–940 (2013).
44. S. Sengupta, K. K. Dey, H. S. Muddana, T. Tabouillot, M. E. Ibele, P. J. Butler and A. Sen, *J. Am. Chem. Soc.* **135**(4), 1406–1414 (2013).
45. G. Zhao, A. Ambrosi and M. Pumera, *Nanoscale* **5**(4), 1319–1324 (2013).
46. M. S. Baker, V. Yadav, A. Sen and S. T. Phillips, *Angew. Chem. Int. Ed.* **52**(39), 10295–10299 (2013).

47. V. Yadav, J. D. Freedman, M. Grinstaff and A. Sen, *Angew. Chem. Int. Ed.* **52**, 10997–11001 (2013).
48. J. L. Anderson, *Annu. Rev. Fluid. Mech.* **21**, 61–99 (1989).
49. J. L. Anderson, *Ann. N.Y. Acad. Sci.* **469**(1), 166–177 (1986).
50. R. P. Feynman, *Engineering and Science* **23**(5), 22–36 (1960).
51. H. C. Berg and D. A. Brown, *Nature* **239**(5374), 500–504 (1972).
52. S. M. Block, *Cell* **93**(1), 5–8 (1998).
53. E. Bonabeau, G. Theraulaz and M. Dorigo, *Swarm Intelligence: From Natural to Artificial Systems* (Oxford University Press, 1999).
54. E. M. Purcell, *Am. J. Phys.* **45**(1), 3–11 (1977).
55. E. Lauga, *Phys. Rev. Lett.* **106**(17), 178101 (2011).
56. R. Eli, *J. Colloid Interface Sci.* **83**(1), 77–81 (1981).
57. R. Golestanian, T. B. Liverpool and A. Ajdari, *New J. Phys.* **9**(5), 126 (2007).
58. S. N. Rasuli and R. Golestanian, *J. Phys-Condens. Mat.* **17**(14), S1171–S1176 (2005).
59. H. R. Jiang, N. Yoshinaga and M. Sano, *Phys. Rev. Lett.* **105**(26), (2010).
60. L. Baraban, R. Streubel, D. Makarov, L. Han, D. Karnaushenko, O. G. Schmidt and G. Cuniberti, *ACS Nano* **7**(2), 1360–1367 (2012).
61. R. Golestanian, T. B. Liverpool and A. Ajdari, *Phys. Rev. Lett.* **94**(22), 220801 (2005).
62. S. Ebbens, M. H. Tu, J. R. Howse and R. Golestanian, *Phys. Rev. E* **85**(2), (2012).
63. D. Okawa, S. J. Pastine, A. Zettl and J. M. J. Fréchet, *J. Am. Chem. Soc.* **131**(15), 5396–5398 (2009).
64. Lauga E. and A. M. J. Davis, *J. Fluid Mech.* **705**, 120–133 (2012).
65. R. Sharma, S. T. Chang and O. D. Velev, *Langmuir* **28**(26), 10128–10135 (2012).
66. H. Zhang, W. Duan, L. Liu and A. Sen, *J. Am. Chem. Soc.* **135**(42), 15734–15737 (2013).
67. J. G. Gibbs and Y. P. Zhao, *Appl. Phys. Lett.* **94**(16), 163104–163103 (2009).
68. A. A. Solovev, Y. Mei, E. Bermúdez Ureña, G. Huang and O. G. Schmidt, *Small* **5**(14), 1688–1692 (2009).
69. S. Sanchez, A. A. Solovev, Y. Mei and O. G. Schmidt, *J. Am. Chem. Soc.* **132**(38), 13144–13145 (2010).
70. W. Gao, S. Sattayasamitsathit, J. Orozco and J. Wang, *J. Am. Chem. Soc.* **133**(31), 11862–11864 (2011).
71. P. Tierno, R. Golestanian, I. Pagonabarraga and F. Sagués, *J. Phys. Chem. B* **112**(51), 16525–16528 (2008).
72. A. Ghosh and P. Fischer, *Nano Lett.* **9**(6), 2243–2245 (2009).
73. W. Gao, S. Sattayasamitsathit, K. M. Manesh, D. Weihs and J. Wang, *J. Am. Chem. Soc.* **132**(41), 14403–14405 (2010).
74. P. Fischer and A. Ghosh, *Nanoscale* **3**(2), 557–563 (2011).
75. S. Tottori, L. Zhang, F. Qiu, K. K. Krawczyk, A. Franco-Obregon and B. J. Nelson, *Adv. Mater.* **24**(6), 811–816 (2012).

76. S. Tottori, L. Zhang, F. Qiu, K. K. Krawczyk, A. Franco-Obregón and B. J. Nelson, *Adv. Mater.* **24**(6), 709–709 (2012).
77. H. Masoud and A. Alexeev, *Soft Matter* **6**(4), 794–799 (2010).
78. E. E. Keaveny, S. W. Walker and M. J. Shelley, *Nano Lett.* **13**(2), 531–537 (2013).
79. W. Wang, L. A. Castro, M. Hoyos and T. E. Mallouk, *ACS Nano* **6**(7), 6122–6132 (2012).
80. S. Ahmed, W. Wang, L. O. Mair, R. D. Fraleigh, S. Li, L. A. Castro, M. Hoyos, T. J. Huang and T. E. Mallouk, *Langmuir* **29**(52), 16113–16118 (2013).
81. V. Garcia-Gradilla, J. Orozco, S. Sattayasamitsathit, F. Soto, F. Kuralay, A. Pourazary, A. Katzenberg, W. Gao, Y. Shen and J. Wang, *ACS Nano* **7**(10), 9232–9240 (2013).
82. M. Ibele, T. E. Mallouk and A. Sen, *Angew. Chem. Int. Ed.* **48**(18), 3308–3312 (2009).
83. W. Duan, R. Liu and A. Sen, *J. Am. Chem. Soc.* **135**(4), 1280–1283 (2013).
84. V. Yadav, H. Zhang, R. Pavlick and A. Sen, *J. Am. Chem. Soc.* **134**(38), 15688–15691 (2012).
85. J. J. McDermott, A. Kar, M. Daher, S. Klara, G. Wang, A. Sen and D. Velegol, *Langmuir* **28**(44), 15491–15497 (2012).
86. M. E. Ibele, P. E. Lammert, V. H. Crespi and A. Sen, *ACS Nano* **4**(8), 4845–4851 (2010).
87. R. Golestanian, *Phys. Rev. Lett.* **108**(3), 038303 (2012).
88. A. A. Solovev, S. Sanchez and O. G. Schmidt, *Nanoscale* **5**(4), 1284–1293 (2013).
89. W. Duan, M. Ibele, R. Liu and A. Sen, *Eur. Phys. J. E* **35**(8), 77–84 (2012).
90. D. Kagan, S. Balasubramanian and J. Wang, *Angew. Chem. Int. Ed.* **50**(2), 503–506 (2011).
91. Y. Gu and D. Li, *J. Colloid Interface Sci.* **226**(2), 328–339 (2000).
92. J. Bird, *Engineering Mathematics* (Newnes, Oxford, 2007), 5th edn.

Chapter 6

PHASE-FIELD DESCRIPTION
OF SUBSTRATE-BASED MOTILITY
OF EUKARYOTIC CELLS

Igor S. Aranson

Materials Science Division, Argonne National Laboratory,
9700 S. Cass Avenue, Argonne, IL 60439, USA

Jakob Löber

Institut für Theoretische Physik, Technische Universität Berlin,
Hardenbergstrasse 36, 10623 Berlin, Germany

Falko Ziebert

Theoretische Physik I, Universität Bayreuth,
95440 Bayreuth, Germany

1. Introduction

Substrate-based cell motility, emerging spontaneously or in response to external cues, is involved in many biological processes, from morphogenesis, wound healing, immune response, to pathologies, especially in cancer growth and metastasis. Cell motility can also be cleverly employed for cell screening and sorting for the purpose of analysis and identification of healthy and malignant cells. Still, a comprehensive understanding of the underlying mechanisms leading to predictive models of the cell responses to environmental changes or external stimuli has not been achieved.

Simple computational models with predictive capabilities will be useful in the context of cell movement. However, casting cell response into a simple mathematical framework is a very complex task: regulatory pathways are intertwined and the motility machinery associated with cell motion is not fully quantified. The current consensus is that the basic processes involved in substrate-based cell motility are: (i) actin protrusion via polymerization at the cell's front (also called the leading edge), (ii) the intermittent formation of adhesion sites for the cell to transfer momentum to the substrate, (iii) and the detachment of adhesion and possibly myosin-driven contraction at the cell's rear.[1,2] Despite the fact that the cell's protrusion is itself a complex biological process,[3,4] it was modeled recently on the scale of the cell by the level set,[5-7] or the phase field[8-13] methods. The phase-field approach tracks the cell's boundary (i.e., the membrane) self-consistently and is very efficient from the computational viewpoint.

This work is a brief overview of a phase-field approach to cell motility based on a minimal phenomenological model proposed by us in a series of works.[9,11,12] Our approach is based on a description of a closed interface (the cell membrane) coupled to a simplified internal dynamics (related to the actin cytoskeleton). The model also incorporates self-consistent deformations of the substrate due to cell movement and coupling with the substrate-dependent adhesion dynamics. Despite the simplifications, the phase-field model captures diverse phenomena for a variety of motile cells migrating on complex natural and engineered heterogeneous substrates, including the abrupt onset of motion, haptotaxis and durotaxis (i.e., drift due to gradients in cell adhesion and substrate stiffness, respectively),[14,15] selective navigation on patterned adhesive substrates,[16] and even recently observed bi-pedal motion of keratocyte cells.[12,17,18]

2. Phase-Field Model

In Refs. 9, 11, 12 we have proposed a minimal model to describe the motion of a closed interface, identifying the cell's membrane, coupled to simplified description of the acto-myosin cyoskeletal dynamics in the cell and self-consistent response of the substrate. The interface is described implicitly by an auxiliary *phase field* $\rho(x, y; t)$, assuming $\rho = 1$ inside the cell and $\rho = 0$ outside. The phase field varies smoothly between 0 and 1, and the interface can be defined by $\max(|\nabla \rho|)$ or by a fixed value of ρ (e.g., $\rho = 1/2$). The simplest phase field implementation of an interface is given by

$$\partial_t \rho = D_\rho \Delta \rho - (1 - \rho)(\delta - \rho)\rho - \mathbf{v} \cdot \nabla \rho. \tag{1}$$

The first term on the r.h.s. determines the width of the interface and the second one is the variational derivative $\frac{\delta F}{\delta \rho}$ of a model "free energy" $F(\rho)$ that has a double well structure with minima at $\rho = 0$ and 1. The value of δ determines which of these two "phases" (the inside, $\rho = 1$, or the outside, $\rho = 0$) is favored. For $\delta = 1/2$ both phases have the same free energy and hence the planar interface is stationary for $\mathbf{v} = 0$. The last term describes the advection of the interface by the external velocity field $\mathbf{v}$. Later the advection velocity $\mathbf{v}$ will be connected to the dynamics of acto-myosin and adhesive properties of the substrate. Note that solving Eq. (1) in one dimension for $\delta = 1/2$ and $\mathbf{v} = (v_0, 0)$, $v_0 = const$, leads to a moving tanh-solution

$$\rho_0(x, t) = \frac{1}{2} \left[\tanh \left(\frac{x - x_0 - v_0 t}{2\sqrt{2D_\rho}} \right) + 1 \right]. \qquad (2)$$

This solution is known from the Ginzburg–Landau theory; here x_0 is the position of the interface connecting the phases 0 and 1. The expression for the front Eq. (2) is also approximately valid for a velocity field $v_0(x)$ varying slowly in space compared to the interface width $1/\sqrt{D_\rho}$. One can generalize the front solution given by Eq. (2) to the case of a circular "droplet" of radius R, large compared to the interface width, i.e., $R \gg 1/\sqrt{D_\rho}$. An equation for the evolution of the droplet's radius R can be obtained asymptotically by using the front solution given by Eq. (2) in polar coordinates (r, θ). Substituting it in the form $\rho(x, y, t) = \rho_0(r - R(t))$ into Eq. (1), and applying the solvability condition with respect to $\partial_r \rho_0$, and using that in polar coordinates the Laplace operator $\partial_r^2 + r^{-1}\partial_r \approx \partial_r^2 + R^{-1}\partial_r$ for the front solution, one obtains that the radius of the droplet R evolves according to

$$\frac{dR}{dt} = -\frac{D_\rho}{R} - \sqrt{2D_\rho}(\delta - 1/2). \qquad (3)$$

Thus, the droplet shrinks for $\delta > 1/2$ and droplets of large radius R spreads for $\delta < 1/2$. This behavior, however, is inconsistent with the fact that cells do not change their area/volume in the course of motion. In order to enforce the volume conservation, the following global constraint can be implemented[9] (note that the choice is not unique,[8,10] used a different implementation of volume conservation)

$$\delta[\rho] = \frac{1}{2} + \mu \left[\iint \rho \, dx \, dy - V_0 \right]. \qquad (4)$$

Here μ is the stiffness of the area constraint and the term in brackets is the difference of the actual area and the prescribed (initial) area V_0. If, for instance, the area is too small, the resulting $\delta < 1/2$ will lead to expansion of the phase $\rho = 1$, i.e., the cell will extend/advance to restore the prescribed V_0. If the area is too big, the resulting $\delta > 1/2$ will force the phase $\rho = 1$ to contract.

3. Coupling to Actin Polymerization Dynamics

Polymerization forces in the cytoskeleton will change the balance between the $\rho = 0, 1$ phases in Eq. (1). The acto-myosin cytoskeleton can be implemented differently because the actin-related motility is manifested differently for various cell types. Most recent approaches to the actin cytoskeleton include the associated flow field (active gel theories,[19] Mogilner and co-workers[20]) or the concentration fields for each constituent of the cytoskeleton (e.g., bundled versus unbundled actin[8] or G-actin monomers and F-actin filaments).

The actin network is described by a vector field, $\mathbf{p}(x, y; t)$: its direction corresponds to the averaged orientation of actin and its absolute value measures the degree of orientation. There are two major factors that determine the most biologically important coupling mechanisms to the cell's membrane. (1) Polymerization of actin filaments that is governed by membrane-associated proteins (nucleators and regulators like WASP and Arp2/3). That mechanism is described by a source term, $\beta\nabla\rho$, in equation for $\mathbf{p}$, Eq. (5). Note that this term favors the orientation of the actin network to be normal to the cell membrane. (2) Thread-milling. Actin filaments, polymerizing at the boundary, push against the membrane due to the ratcheting of added monomers.[21] This action of cytoskeleton on the cell membrane will be incorporated into the model by setting the advective velocity $\mathbf{v}$ in Eq. (1) to be proportional to the local polarization of the acting network, i.e., $\mathbf{v} = \alpha\mathbf{p}$. The factor α, in turn, can depend on adhesive and frictional properties of the substrate. The equation for the polarization of actin network $\mathbf{p}$ can be cast in the form

$$\partial_t \mathbf{p} = D_p \Delta\mathbf{p} - \beta\nabla\rho - \tau_1^{-1}\mathbf{p} - \tau_2^{-1}(1 - \rho^2)\mathbf{p}. \tag{5}$$

This equation also includes degradation of actin (with rate τ_1) and a diffusive/elastic-like term $D_p\Delta\mathbf{p}$. Since a value of $\mathbf{p}$ outside the cell will not change the dynamics, the last term explicitly suppresses $\mathbf{p}$ outside the cell.

Consider first a center-symmetric cell. Since the actin network is created normal to the cell membrane, the actin continuously pushes but is held back by the volume conservation, i.e., the cell cannot move. In order to sustain a moving state, the symmetry must be broken. It has been shown in Ref. 22 that for keratocyte cells myosin motors are responsible for the symmetry breaking. Two effects are identified: first, motors induce local contraction. Second, motors facilitate the formation of acto-myosin bundles at the rear of the cell.

Correspondingly, the contraction effect can be modeled by adding a term $-\sigma|\mathbf{p}|^2$ to Eq. (4) for $\delta[\rho]$. Following the ideas formulated in active gel theories,[19] this term describes actin contraction mediated by myosin motors, with an associated rate parameter σ. The second effect describes myosin motor mediated formation of bundles at the rear, and can be modeled by adding a symmetry-breaking term of the form $-\gamma[(\nabla\rho)\cdot\mathbf{p}]\mathbf{p}$ to Eq. (5). This term, breaking the $\pm\mathbf{p}$ reflection symmetry, can be derived from simple motor dynamics.[9] It describes an increased motor activity at the rear and suppression of polarization $\mathbf{p}$ by formation of anti-parallel (nematic-like) bundles. Note that the polar order parameter, described by the vector $\mathbf{p}$, is reduced for anti-parallel, contractile bundles.

4. Dynamics of Adhesion

Cell adhesion is a complex multi-stage process. It involves interactions of several proteins forming complexes linking the internal actin cytoskeleton to the extracellular matrix (ECM). Additional complications come from the fact that the system is mechanosensitive. Namely, the formation of adhesion sites depends on forces, both generated by the cell in the course of motion or applied externally. Adhesion dynamics can also be time-dependent: adhesion sites undergo maturation.[23,24] In the following, we consider the simplest case of motility-related adhesion. We assume that rectified actin polymerization creates a pushing force close to the membrane, propelling the cell forward *only if* it locally adheres. Thus, this adhesion-mediated force transfer to the substrate is consistent with the observations by traction force measurements for both spreading and crawling cells.[25,26]

In order to include basic features of adhesion dynamics, we let the propulsion term, i.e., the advective term $\alpha\mathbf{p}\cdot(\nabla\rho)$ in Eq. (1), depend on the local density of adhesive bonds, $A(x,y;t)$. We assume a linear dependence $\alpha \to \alpha(A) \simeq \alpha_0 A$ — an increase in the amount of adhesive bonds directly increases the force of propulsion. However, experiments show that for too strong adhesion, the cell speed decreases again.[27] It implies that breaking of an adhesive bond takes too much energy. This effect can be captured by a nonlinear (non-monotonic) function $\alpha(A)$. The equation for the density of adhesive bonds[11] (cf. also[10] for a discrete implementation of adhesion) can be cast in the form

$$\partial_t A = D_A \Delta A + \rho\big(a_0 p^2 + a_{nl} A^2\big) - \big(d(u) + sA^2\big)A \,. \tag{6}$$

The terms on the r.h.s. of Eq. (6) describe diffusion, attachment and detachment/limitation of adhesive bonds correspondingly. We employed simple implementation rules: formation of adhesive bonds is restricted to the interior of the cell, which is provided by a common factor of ρ. The attachment term has two contributions: a linear term — proportional to p^2, guarantees the presence of actin. The second, nonlinear term, accounts for several mechanisms facilitating attachment of additional bonds if a bond has been already formed (for instance, locally reduced membrane fluctuations). The detachment rate is based on the following mechanisms: the term $d(u)A$ describes breaking of the bonds due to the substrate displacement $u(x, y; t)$ (or the elastic restoring force). The cubic term sA^3 saturates the local density of A and is the simplest implementation of an excluded volume constraint.

5. Response of the Substrate

In the course of motion, the cell exerts forces on the substrate; in turn, the substrate deformations affect the adhesion. An equation for the local substrate displacement $u(x, y; t)$ can be derived in thin layer approximation from linear elasticity theory. For deformations in a thin visco-elastic layer of thickness h attached to a rigid substrate one obtains[12]

$$\partial_t u = -\frac{1}{\eta}\left(Gu - \frac{1}{\xi}\left(T + O(h)\right)\right). \tag{7}$$

Here G is the elastic modulus of the substrate, η describes viscous relaxation in the adhesive layer and $T = -\xi A\rho\left(p + \frac{\langle A p \rho\rangle}{\langle A\rho\rangle}\right)$ is the traction force exerted by the cell. The traction is proportional to both ρ (restriction to the cell interior) and adhesive bond density A. The first traction contribution is opposite to the local propulsion direction p, the second contribution is due to friction with the substrate. ξ describes the efficiency of force transmission.[12] Correspondingly, the total traction $\langle T\rangle = \int T dx\, dy = 0$ is zero, as it should be for a force-free self-propelled object. One can further simplify the system by a projection of the deformation field on the force dipole in the direction of motion, leading to an effective equation for an overall (cell-averaged) substrate displacement U like $\frac{d}{dt}U = -\frac{1}{\eta}(GU + V)$.[11] While this replacement of the two-dimensional displacement field u by a single one-dimensional spring U in the direction of motion is adequate for many situations, like motion of cell on patterned adhesive substrates, one has to consider the full Eq. (7) for the description of durotaxis and bi-pedal motion.

5.1. *Variety of dynamic states*

To explore a variety of dynamic states, equations of motion (1), (5), (6), (7) were solved numerically in a double-periodic square domain, using a quasi-spectral Fourier method implemented on GPUs. In most cases we used

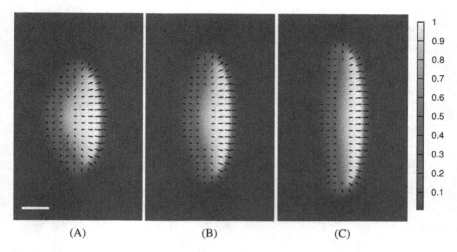

Fig. 1.   Evolution of shapes for a steadily-moving cell for an increasing contractility parameter σ. Colors encode the phase-field $\rho(x, y)$ from blue (minimum) to yellow (maximum); arrows display the actin polarization **p**. Cell moves to the right.

512×512 FFT harmonics, and calculations were performed with double precision.

A multiplicity of dynamic states, successfully reproduced by the model, is shown in Figs. 1–4. Figure 1 illustrates the evolution of cell shape as a function of the contractility parameter σ. Increasing this parameter from 0 towards a value of order one ($\sigma \sim O(1)$) results in an overall increase of the aspect ratio of the cells. Thus, the situation of small σ roughly corresponds to fibroblast-like cells characterized by a triangular shape, whereas large-σ cells resemble crescent-like keratocytes.

The model also captures the occurrence of stick-slip motion for cells moving on soft deformable substrates, as illustrated in Fig. 2. The following generic mechanism of stick-slip was suggested in Ref. 11. The cell speeds up when the adhesion sites are forming. Since the adherent cell exerts dipolar forces on the substrate, the substrate deformation increases in its absolute value. If the stiffness of the substrate is low enough, the substrate deformation exceeds the critical value, and the adhesive contacts rapidly break. However, the cell has still to slow down and adjust its shape to the new conditions. Once the substrate relaxes, new adhesion sites are allowed to form again and the cycle restarts.

Figure 3 illustrates motion on a patterned adhesive substrate. A similar situation was studied experimentally for keratocytes in Ref. 16, using micro-contact printing of fibronectin for regions of high adhesiveness and of

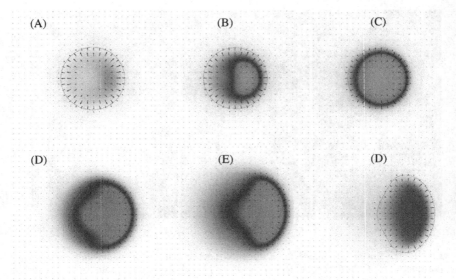

Fig. 2. Sequence of snapshots illustrating evolution of cell shapes in the course of a stick-slip cycle for 6 consecutive moments of time: (A) $t = 260$, (B) $t = 266$, (C) $t = 270$, (D) $t = 280$, (E) $t = 290$, (F) $t = 294$. The cell moves to the right. The cell boundary is depicted by the green line, black arrows show the local actin filament orientation. The density of adhesion sites is color coded (with white corresponding to $A = 0$, blue to $A = 0.5$ and red to $A = 1$).

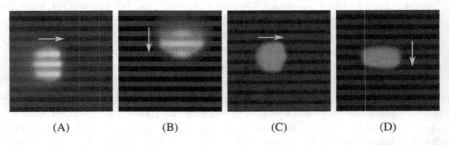

Fig. 3. (A) Motion of a cell on a rigid substrate ($G = 0.2$) with alternating stripes of high adhesiveness parameter $a_0 = 0.15$ (blue) and no adhesiveness ($a_0 = 0$ for black stripes). The cell positions itself symmetrically and moves parallel to the stripes in a steady fashion. (B) Motion of a cell on a substrate with $G = 0.1$, and with alternating stripes of low adhesiveness parameter $a_0 = 0.0015$ (blue) and $a_0 = 0$ (black). After moving initially along the stripes, the cell turns and moves perpendicular to the stripes in a stick-slip fashion. (C) Trajectories of cells on substrates with two different width ratios of adhesive to non-adhesive stripes. For large widths of the adhesive stripe, the cell moves along the stripes. (D) A gradual decrease in the stripe width forces the cell to move perpendicular to stripes.

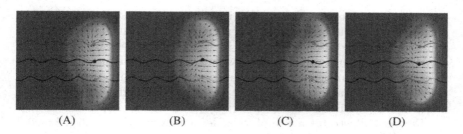

(A)   (B)   (C)   (D)

Fig. 4.   Illustration of bipedal motion. (A)–(D) show snapshots including the displacement field. Parameters are $G = 0.2$, $\alpha = 4.25$. The green/red curves in the lowest panel are $r_0 + r_\pm$. Here r_0 is the cell's center of mass (c.o.m), and $r_\pm$ are c.o.m.s of the upper ($+$) and respectively lower ($-$) half of the cell.

poly-L-lysine-PEG block copolymers for practically non-adhesive regions. The experiments revealed that the cells have a tendency to move parallel to the adhesive stripes, while the cell's centroid is positioned in the middle of the stripe. In our model, the substrate's selective adhesiveness is modeled by a spatial modulation of the parameter a_0 corresponding to the rate of adhesion complex formation. For stripes with large values of parameter a_0 we observed that the cell positions itself symmetrically with respect to the stripes, see Fig. 3A. Thus, the cell's center-of-mass migrates towards the center of the high adhesiveness stripe, and the cell moves along the stripes. This behavior is in faithful agreement with the experimental observation in Ref. 16.

A fundamentally different behavior was obtained for cells moving on striped substrates with lower values of the adhesion parameter a_0, as it is illustrated in Fig. 3B. In this regime, the homogeneous system displays stick-slip motion. The displayed cell was stimulated to move along the stripes by initial conditions. However, after some time, the cell slows down, abruptly changes the direction, spreads along the stripe in order to maximize the contacts with the high-adhesiveness region, and begins to move perpendicular to the stripes. We also observed that in this regime the cell may randomly reverse the direction. Thus, our model makes a nontrivial prediction on a new type of motion on patterned low adhesive substrates.

It is rather difficult to fabricate surfaces with a significant modulation of the adhesive strength. However, it is much simpler to vary the relative width of adhesive/non-adhesive stripes, while keeping the period of the pattern fixed.[16] We investigated this situation in our model and observed a behavior similar to that shown in Figs. 3A and 3B. Namely, for large widths of the adhesive stripes, the cell moves along the stripes in agreement with previous simulations, Fig. 3C. On gradually decreasing the width of the adhesive stripes, we observed an instability: the cell exhibits a kind of rocking motion, and eventually turns perpendicular to the stripes, Fig. 3D. For very small widths reverse trend is observed: the cell stretches in the direction of motion in order to fit between two stripes, and moves along the stripes. These simulations results, signifying the sensitivity of cell's motion to the commensurability between the size of the cell and the period of the modulation, constitute another nontrivial prediction for experiments.

Figure 4 illustrates a complex mode of movement, so-called bipedal motion,[17,18] occurring for cells moving on a soft substrate. In contrast to the stick-slip motion shown in Fig. 2, in the course of bipedal motion the cell exhibits periodic lateral out-of-phase oscillations while preserving the overall propagation direction. For slightly different parameters, the bipedal motion itself becomes unstable, and the cell veers off the straight path and follows a slightly curved trajectory accompanied by asymmetric shape oscillations (wandering bipedal).[12] In order to capture the bipedal states and durotaxis, we resolved the local substrate displacement and traction as described by Eq. (7). The inhomogeneously distributed force affects the adhesion and provides a feedback on the overall shape and motion. In contrast to Ref. 11 where the substrate response was modeled by a single one-dimensional spring, new dynamic states emerge at the boundary of the transition from stick-slip to steady motion due to the local nature of the rupture of the adhesive contacts.

6. Conclusions

In this brief review we have discussed a phase-field approach to model cell motion. The phase field description for the cell's moving boundary circumvents a major (numerical, but also conceptional) bottleneck associated with the tracking of moving interfaces. The approach successfully reproduces a variety of experimental observations on cell movement and leads to testable predictions. The approach can be extended step by step in order to increase the level of detail, and, furthermore, can be eventually adopted to different phenotypes and specific cell types (keratocytes, fibroblast, neutrophils, see Ref. 28).

The phase-field approach could bridge the gap between simple one-dimensional or fixed shape models[29,30] and more complex, partially three-dimensional, approaches that have been started to be developed but do not allow a very detailed study due to their complexity.[6,31] On the other hand, some generic features that are currently established by even more coarse-grained models of soft self-propelled particles, cf., Refs. 32 and 33, could be validated in the specific context of motile cells.

Cell motility is an active field of research, and the literature on cell motility modeling is rapidly expanding. Correspondingly, various groups employ different methods to treat similar phenomena, see for recent reviews Refs. 34–36. A conceptual simplicity of the phase-field approach might help to achieve important comparisons and establish a unified framework for modeling the motion of different cell types, as well as of related biomimetic objects (see e.g., Ref. 37). The phase field approach can be extended to multiple cells, and can also facilitate the description of motion of cell layers, as occurring e.g., during development or wound closing.[38]

Acknowledgments

J. L. acknowledges financial support by the DFG via GRK 1558. F. Z. thanks the DFG for partial support via IRTG 1642 Soft Matter Science. Work by I.S.A was supported by the US Department of Energy (DOE), Office of Science, Basic Energy Sciences (BES), Materials Science and Engineering Division.

References

1. M. Abercrombie, *Proc. R. Soc. London B* **207**, 129 (1980).
2. M. Sheetz, D. Felsenfeld and C. Galbraith, *Trends Cell Biol.* **8**, 51 (1998).
3. T. D. Pollard, *Nature* **422**, 741 (2003).
4. T. D. Pollard and G. G. Borisy, *Cell* **112**, 453 (2003).
5. M. Machacek and G. Danuser, *Biophys. J.* **90**, 1439 (2006).
6. E. Kuusela and W. Alt, *J. Math. Biol.* **58**, 135 (2009).
7. C. Shi, C.-H. Huang, P. Devreotes and P. Iglesias, *PLoS Cell Biol.* **9**, e1003122 (2013).
8. D. Shao, W. J. Rappel and H. Levine, *Phys. Rev. Lett.* **105**, 108104 (2010).
9. F. Ziebert, S. Swaminathan and I. S. Aranson, *J. R. Soc. Interface* **9**, 1084 (2012).
10. D. Shao, H. Levine and W.-J. Rappel, *Proc. Natl. Acad. Sci. USA* **109**, 6851 (2012).
11. F. Ziebert and I. S. Aranson, *PLOS ONE* **8**, e64511 (2013).
12. J. Löber, F. Ziebert and I. S. Aranson, *Soft Matt* **10**, 1365 (2014).
13. W. Marth and A. Voigt, Signaling networks and cell motility: A computational approach using a phase field description, *Journal of Mathematical Biology* (2013), pp. 1–22.
14. C. G. Rolli, H. Nakayama, K. Yamaguchi, J. Spatz, R. Kemkemer and J. Nakanishi, *Biomaterials* **33**, 2409 (2012).
15. L. Trichet, J. Le Digabel, R. Hawkins, S. R. Vedula, M. Gupta, C. Ribrault, P. Hersen, R. Voituriez and B. Ladoux, *Proc. Natl. Acad. Sci. USA* **109**, 6933 (2012).

16. G. Csucs, K. Quirin and G. Danuser, *Cell Motil. Cytoskeleton* **64**, 11, 856–867 (2007).
17. E. Barnhart, G. Allen, F. Jülicher and J. Theriot, *Biophys. J.* **98**, 6, 933–942 (2010).
18. A. J. Loosley and J. X. Tang, *Phys. Rev. E* **86**, 031908 (2012), doi:10.1103/PhysRevE.86.031908.
19. K. Kruse, J. F. Joanny, F. Jülicher, J. Prost and K. Sekimoto, *Eur. Phys. J. E* **16**, 5 (2005).
20. B. Rubinstein, M. F. Fournier, K. Jacobson, A. B. Verkhovsky and A. Mogilner, *Biophys. J.* **97**, 1853 (2009).
21. C. S. Peskin, G. M. Odell and G. F. Oster, *Biophys. J.* **65**, 316 (1993).
22. P. T. Yam, C. A. Wilson, L. Ji, B. Hebert, E. L. Barnhart, N. A. Dye, P. W. Wiseman, G. Danuser and J. A. Theriot, *J. Cell Biol.* **178**, 1207 (2007).
23. K. Burridge and M. Chrzanowska-Wodnicka, *Annu. Rev. Cell Dev. Biol.* **12**, 463 (1996).
24. U. S. Schwarz and M. L. Gardel, *J. Cell Sci.* **125**, 3051 (2012).
25. M. Dembo and Y. L. Wang, *Biophys. J.* **76**, 2307 (1999).
26. M. F. Fournier, R. Sauser, D. Ambrosi, J.-J. Meister and A. B. Verkhovsky, *J. Cell Biol.* **188**, 287 (2010).
27. S. P. Palecek, J. C. Loftus, M. H. Ginsberg, D. A. Lauffenburger and A. F. Horwitz, *Nature* **385**, 537 (1997).
28. S. J. Henry, J. C. Crocker and D. A. Hammer, *Integrative Biology* **6**, 348 (2014).
29. P. Recho, T. Putelat and L. Truskinovsky, *Phys. Rev. Lett.* **111**(10), 108102 (2013).
30. C. Blanch-Mercader and J. Casademunt, *Phys. Rev. Lett.* **110**, 7, 078102 (2013).
31. W. Alt and M. Dembo, *Math. Biosci.* **156**, 207 (1999).
32. T. Ohta and T. Ohkuma, *Phys. Rev. Lett.* **102**, 154101 (2009).
33. M. Tarama, Y. Itino, A. Menzel and T. Ohta, *Eur. Phys. J. Spec. Top.* **223**(1), 121–139 (2014).
34. W. R. Holmes and L. Edelstein-Keshet, *PLoS Comput. Biol.* **8**, e1002793 (2012).
35. U. S. Schwarz and S. A. Safran, *Rev. Mod. Phys.* **85**(3), 1327 (2013).
36. G. Danuser, J. Allard and A. Mogilner, *Annu. Rev. Cell Dev. Biol.* **29**, 1 (2013).
37. G. V. Kolmakov, A. Schaefer, I. Aranson and A. C. Balazs, *Soft Matter* **8**, 180 (2012).
38. M. H. Köpf and L. M. Pismen, *Soft Matter* **9**, 3727 (2013).

Chapter 7

FROM COLLOID THERMOPHORESIS TO THERMOPHORETIC MACHINES

Marisol Ripoll[*,‡] and Mingcheng Yang[*,†,§]

*Theoretical Soft-Matter and Biophysics,
Institute of Complex Systems, Forschungszentrum Jülich,
52425 Jülich, Germany

†Beijing National Laboratory for Condensed Matter Physics and
Key Laboratory of Soft Matter Physics, Institute of Physics,
Chinese Academy of Sciences, Beijing 100190, China
‡m.ripoll@fz-juelich.de
§mcyang@iphy.ac.cn

1. Introduction

A temperature gradient applied to fluid mixtures induces two effects. Heat conduction refers to the transport of energy from the warm to the cold areas, which occurs on a very fast time scale. Thermodiffusion, thermal diffusion, or Soret effect refers to the transport of mass induced by a temperature gradient, which occurs on a longer time scale than heat conduction. Thermodiffusion frequently translates into component segregation, which is dictated by the microscopic interactions within the system, and very importantly by the interactions between the components. This segregation has been investigated in gas mixtures, molecular liquid mixtures, and also in macromolecular systems. When a temperature gradient is applied to a colloid in solution, a measurable particle drift is induced towards the cold or the warm areas. Thermodiffusion of colloidal or polymeric solutions is commonly referred to as *thermophoresis*.

Thermodiffusion was first observed more than 150 years ago[1,2] in salt water mixtures, where the salt was observed to accumulate in the cold side. Since then, it has found a large number of applications like isotope separation,[3,4] crude oil characterization,[5] or separation of macromolecules in solution.[6] In the last two decades, new and precise experimental techniques have stimulated extensive research interest to investigate this effect in various systems,[7,8] examples of these techniques are thermal diffusion cells, thermogravitational columns, or thermal diffusion forced Rayleigh scattering. Furthermore, precise techniques to control the temperature at microscales are allowing the development of microfluidic applications.[9,10] In recent studies, the interaction of proteins in biological liquids has been analyzed in terms of their thermophoretic behavior,[11] which is currently being used for their identification. Due to the frequent thermal gradients in the early ocean, thermodiffusion has also been found to be a facilitating agent for the origin of life.[12,13]

2. Thermodiffusion

In a two component mixture thermodiffusion can be characterized phenomenologically by the quantification of the particle flux J of one of the components in the direction of a temperature gradient ∇T,[14]

$$J = -\tilde{n} D_m \nabla x - \tilde{n} D_T x (1 - x) \nabla T, \tag{1}$$

where $\tilde{n} = n + n'$ is the average total number density, with n and n' the density of the two components, $x = n/\tilde{n}$ the molar fraction of component one, D_m is the mutual diffusion coefficient, and D_T the thermal diffusion coefficient. The so-called *Soret coefficient* is defined as the ratio of the two diffusion coefficients

$$S_T \equiv \frac{D_T}{D_m}, \tag{2}$$

and indicates how strongly the two components separate. In the stationary state, the particle flux vanishes, and the Soret coefficient can be obtained from the molar fraction and temperature distributions

$$S_T = -\frac{1}{x(1 - x)} \frac{\nabla x}{\nabla T}. \tag{3}$$

Note that by convention, a positive S_T indicates that the relative accumulation of component one is higher in the cold side, while a negative S_T will display a reverse behavior. Equation (3) is a very well-established expression in the community and constitutes the standard method to quantify the thermodiffusion phenomena in concentrated mixtures. Nevertheless, it does not reveal the physical origin of the separation between components, and it is of little use in the investigation of dilute systems.

Alternatively to Eq. (1), the particle flux in a temperature gradient can be obtained by the Fokker–Planck description of Brownian motion in inhomogeneous suspensions,[15,16]

$$\mathbf{J}(\mathbf{r}) = n(\mathbf{r})\mu(\mathbf{r})\mathbf{f} - \mu(\mathbf{r})\nabla[n(\mathbf{r})k_B T(\mathbf{r})], \qquad (4)$$

with $\mathbf{f}$ the mechanical driving force and $\mu(\mathbf{r})$ the effective mobility of a single particle in a background solvent. The force $\mathbf{f}$ was originally understood just as the externally applied forces, but the concept was in an early stage extended to consider also the mechanical driving forces exerted on single particles by an inhomogeneous surrounding fluid,[17] which in the case of a temperature gradient corresponds to the *thermophoretic force* (see sketch in Fig. 1). By assuming that Eq. (4) is valid for the two components of a mixture, and comparing with Eq. (3), an alternative expression of the Soret coefficient has been proposed in terms of the difference of mechanical driving forces exerted on the particles,[18]

$$S_T = (\mathbf{f}' - \mathbf{f}) \frac{1}{k_B \bar{T} \nabla T}, \qquad (5)$$

with $\bar{T}$ the average temperature. This alternative expression can be employed in the determination of the colloidal thermophoretic properties by determining the thermophoretic forces. These are directly accessible in simulations and can be estimated by analytical arguments. In some systems, they can also be experimentally obtained with techniques such as single-particle tracking.

2.1. *Colloid thermophoresis*

Thermodiffusion of colloids has been extensively investigated by means of various experimental techniques,[19–22] different analytical approaches,[23–27] and recently also by means of simulations.[28,29] Most colloids in solution show a *thermophobic* behavior, this is, they tend to migrate to cold areas, although examples of *thermophilic* colloids, which tend to migrate to warm areas, can also be found.[8,30] The particularities of the interactions between the colloid and the surrounding solvent have shown to determine their thermophoretic behavior. Properties that strongly influence the related thermophoretic forces are for example the colloidal interfacial properties, solvent polarity, colloidal charge, average temperature, or particle size. Moreover, systems with varying colloidal concentration have shown that also colloid–colloid interactions can contribute significantly to the thermo-diffusive properties of the solution.[31–33]

In the case of components with a large size separation, the thermophoretic forces exerted on each component are very different. The force

Fig. 1. The thermophoretic force on a colloid is not externally applied but results from the inhomogeneous interactions with the surrounding solvent. The colloid will be thermophobic when it drifts to the cold areas, or thermophilic when it drifts to the warm areas.

on the small component $\mathbf{f}'$ in Eq. (5) can be typically neglected. This is the case for most colloidal dispersions and polymer solutions. For a colloid, or a macromolecule in solution, the thermophoretic force $\mathbf{f}_T$ can be understood as resulting from the imbalance of the solvent interactions which drive the particle to the cold or to the warm region (see sketch in Fig. 1). This force is directly proportional to the temperature gradient as deduced from Eq. (5)

$$\mathbf{f}_T = -\alpha_T k_B \nabla T, \tag{6}$$

with $\alpha_T = \bar{T} S_T$, the thermal diffusion factor.

So far no general theoretical explanation is found that can precisely predict the direction and strength of the migration of the colloids along the temperature gradient. The size of the colloid, for example, is know to play an important role, but it is still under debate if the increase of the Soret coefficient is linear or quadratic with the particle radius.[20–22] Similarly, the Soret coefficient typically increases with the average temperature, although there are examples of the opposite behavior. In charged systems, properties like the ionic strength and the valency of the ions are of importance. Structural changes of the surrounding solvent, as induced by the ionic double layer, translate into a change in the thermodiffusive behavior.[26]

2.2. *Simulation models for thermodiffusion*

Besides the varied and powerful experimental techniques which have been employed to investigate thermodiffusion, computer simulations have become important by bringing a new perspective to the understanding and exploitation of the phenomenon. Molecular dynamics simulations (MD)[34,35] identified a first group of general trends of the thermal diffusion behavior of molecular mixtures which are related to the particle properties. Simulations of mixtures of Lennard-Jones particles[36] have shown that: (i) in mixtures differing only in the strength of the attraction potential, the component with stronger attraction accumulates on the cold side, (ii) if the components in the mixture differ only in the mass, the heavier species diffuses to the cooler region, (iii) if the species only differ in size, the smaller component diffuses towards the cooler region. Furthermore, it has been shown that the momentum of inertia of the particles[37–39] can display a significant contribution in the thermodiffusion properties. Specific interactions like hydrogen bonding in aqueous solutions have been shown to play an important role leading even to sign changes in the Soret coefficient, or molecular orientation.[40,41] Lattice simulations of binary systems[42,43] showed that if cross-interactions are stronger than pure interactions, a sign change can be induced.

Until very recently, simulations of colloids in temperature gradients[28,44,45] had been performed with MD of both the colloidal and the solvent particle. Nevertheless, the typical experimental sizes of colloids and solvent particles can be separated by several orders of magnitude, which largely hinders the MD simulations accounting explicitly for the solvent particles. Alternatively, several methods to coarse-grain the solvent dynamics have been developed in the last decades. Relevant examples are lattice Boltzmann (LB)[46] or dissipative particle dynamics (DPD).[47] Both these methods are isothermal in their most extended implementations, although corresponding modifications to these models have been proposed in order to be able to sustain temperature inhomogeneities both for LB[48,49] and DPD.[50–52]

In our work, we mostly employ a hybrid approach that considers the solvent by a relatively new coarse-grained method known as multiparticle collision dynamics (MPC),[53] or stochastic rotation dynamics (SRD).[54,55] This technique has shown to properly include thermal fluctuations, hydrodynamic interactions,[56–59] and also to be able to sustain temperature gradients.[60] Colloidal dynamics are simulated by MD, where the interactions between colloids and/or colloids and solvent are treated by explicit potentials.[29,61] Attractive colloid-solvent interactions have shown to induce thermophobic colloid behavior, while repulsive interactions result into thermophilic colloids.[29] Variations in the softness of the potential, or the average temperature also change the strength of the thermophoretic force. Apart from the explicit and efficient implementation of the non-isothermal solvent, the employed MPC method offers us the possibility of tuning the solvent colloid interactions which turns out to be a very useful practical tool.

3. Self-Propelled Thermophoretic Structures

Molecular motors are ubiquitous in biology, with examples ranging from motor proteins moving along filaments,[62] to swimming bacteria.[63] In such systems self-propulsion is mostly achieved by using chemical energy released from ATP hydrolysis.[64] Recently, synthetic microscale and nanoscale motors have attracted considerable attention due to their potential practical applications and related open theoretical questions.[65,66] A relatively simple and effective strategy to design artificial micromotors has already been found by employing phoretic effects. Phoresis refers not only to thermophoresis, but also to diffusiophoresis or electrophoresis, where the particle drift is caused by gradients of concentration or electric potential. These gradients are frequently a consequence of external constrains, but interestingly, in case one particle is able to produce a local gradient field by itself, self-propulsion may occur. The theoretical basis of this phenomenon has been discussed by different authors.[67,68] Following this line and by means of simulations[69,70] and experiments,[71,72] a chemical reaction catalyzed asymmetrically on particle's surface has shown to translate into a diffusiophoresis motor. More recently, a Janus particle has shown to display self-propelled motion due to thermophoresis.[73] In their experiments, Jiang *et al.* use a half-metal coated colloidal sphere and heat it with a defocused laser. The larger heat absorption of the metal side produces a temperature gradient on the non-metal side, which translates into a self-propelled motion.

3.1. *Swimming microdimer*

Simulations of microdimers that swim due to thermophoresis at low Reynolds numbers are performed by considering two strongly bonded monomers immersed in a hydrodynamic solvent,[74] as depicted in Fig. 2A. The heated bead (h), has a temperature higher than the surrounding fluid. This accounts for a monomer of material as gold that absorbs heat for example from a laser. If the average temperature $\bar{T}$ of the system is kept constant, the surrounding solvent will sustain a steady temperature gradient. The imposed spherical symmetry of the temperature distribution around the h-bead, together with the energy conservation, and the Fourier's law imply a temperature field $T(r) \sim 1/r$ with r the distance to the h-bead center. The non-heated, propelling or phoretic monomer bead (p) is then subject to temperature gradient of the surrounding solvent, which generates a thermophoretic force $\mathbf{f}_T$ on its surface. This translates into a directed motion of the microdimer along the bond direction.

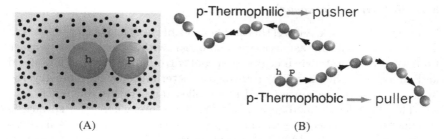

(A)   (B)

Fig. 2. (A) Schematic diagram of the thermophoretic microdimer solution. The h-bead is heated to a fixed temperature and strongly linked to the p-bead. The system has then a radially symmetric temperature gradient. (B) The dimer is self-propelled along the bond direction with a direction motion determined by the thermophoretic properties of the p-bead.

During its motion, the microdimer axis is not fixed due to thermal fluctuation, such that it will freely rotate. This implies that the self-propelled motion will change orientation, giving rise to an enhanced diffusive behavior.[71,73] The swimming behavior of the microdimer can be characterized by v_p, the directed velocity of the dimer along its bond direction. The microdimer will have an average steady velocity when the related viscous force balances the thermophoretic force, $v_p = \mu_p f_T$, where μ_p corresponds to the mobility along the bond direction. Note that there is no thermophoretic force exerted on the h-bead due to symmetric temperature distribution around it. Considering the expression of the thermophoretic force in Eq. (6), the self-propelled velocity can then be expressed as a function of the material properties,

$$v_p = -\mu_p \alpha_T \nabla k_B T. \tag{7}$$

The validity of this relation is general and has been checked in our simulations, since the involved quantities can be independently computed.

Interestingly, the qualitative behavior of the microswimmer can be tuned by changing the material properties. In case the p-bead is thermophilic, it tends to go to higher temperatures, and therefore towards the h-bead, as sketched in Fig. 2B. It can be shown that the thermophilic microdimer behaves as a *pusher*, namely inducing a hydrodynamic lateral attraction.[75] Conversely, in case the p-bead is thermophobic, the thrust is exerted in the direction opposite to the h-bead, and the microdimer behaves as a *puller*, this is inducing a lateral hydrodynamic repulsion.

3.2. *Rotating microgear*

In a very recent work, a novel thermophoretic machine has been proposed.[76] The idea is that an asymmetric microgear with homogeneous surface properties will rotate when heated in a cool surrounding solvent. The speed and direction of the microgear rotation are determined by its geometry, the interactions with the solvent, and the applied temperature differences.

The considered microgear is a solid structure where the surface is a sequence of saw-teeth, surrounded by a solvent, and confined inside a closed container (Fig. 3A). The temperature of the microgear T_g can be first thought of as uniformly fixed by a external constraint, similarly to the temperature of the external wall T_w. When imposing a temperature difference $\Delta T = T_g - T_w$, a steady-state temperature distribution is quickly established in the solvent (Fig. 3B). The environment of the solvent particles close to the summit and the cleft of each gear tooth is quite different due to the different size of the heating areas. The solvent temperature is then different in both positions and varies along the edges, generating a temperature gradient along both the long and the short tooth edges, see ∇T_l and ∇T_s in Fig. 3C.

We have discussed how a colloid in a solvent with a temperature gradient experiences the effect of a thermophoretic force. Similarly, a planar wall in contact with a solvent at varying temperature will also experience a thermophoretic force parallel to the wall. Thus, for a heated microgear ∇T_l and ∇T_s will respectively result in the thermophoretic forces on the

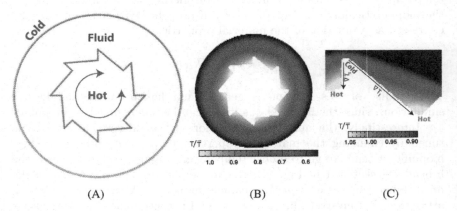

(A)   (B)   (C)

Fig. 3. (A) Schematic setup of a hot self-propelled microgear as employed in the simulations. (B) Solvent temperature distribution. (C) Zoom in the neighbourhood of one microgear tooth, where the temperature gradients along the long and the short edges can be clearly observed.

long edges $f_{T,l}$ and the short edges $f_{T,s}$, parallel to the edges. Depending on the gear-solvent interactions,[29] the thermophoretic forces can be along (thermophilic) or against (thermophobic) the temperature gradient. Due to the asymmetric gear geometry, the thermophoretic forces exert a non-vanishing torque on the gear which results in its unidirectional rotation,

$$\mathcal{T} = \Sigma(\mathbf{R}_l \times \mathbf{f}_{T,l} + \mathbf{R}_s \times \mathbf{f}_{T,s}). \tag{8}$$

Here, $\mathbf{R}_l$ and $\mathbf{R}_s$ respectively refer to the coordinates of the center of force on the long and short edges, and the summation accounts for multiple teeth. It is simple and efficient to consider that the short tooth edge (hence $\mathbf{R}_s$) and long tooth edge are, parallel and perpendicular to the radial direction, respectively. This means that only the forces applied on the long edges contribute to the total torque, and $\mathcal{T} \simeq N R_l f_{T,l}$, with N the number of gear teeth. A hot gear ($\Delta T > 0$) built of a thermophilic material will then rotate with the long teeth edges forward, which in the geometry of Fig. 3 is clockwise. Rotation in the opposite direction is then expected in the case of a cold microgear ($\Delta T < 0$), or when the material is thermophobic. For a gear in equilibrium ($\Delta T = 0$) only pressure forces will be present, such that no net rotation is predicted. Note that for a microgear with symmetric teeth the thermophoretic forces along both sides of each tooth are symmetric with respect to the microgear radial direction, which results in a vanishing torque and vanishing net rotation.

The angular velocity of the gear can be related to the material properties by considering $\omega = \mu_r \mathcal{T}$, with μ_r the microgear rotational mobility. Considering the expression of the thermophoretic force in Eq. (6), and the expression of the torque

$$\omega = -\mu_r N R_l \alpha_T k_B \nabla T_l. \tag{9}$$

The angular velocity is directly proportional to the temperature gradient along the long tooth edges, and to the material thermal diffusion factor α_T. The rotational mobility of a three-dimensional microgear is calculated as $\mu_r = 1/(4\pi\eta R_H^2 h_H)$,[77] with h_H the gear thickness, and R_H the gear hydrodynamic radius.

The model discussed here enforces that the microgear has a constant temperature. Experimentally this corresponds to a microgear fabricated with a material of thermal conductivity much higher than that of the solvent, as it would be the case of a metal or a metal-coated microgear in water solution. Alternatively, thermophoretic microgears can be constructed with materials of low or moderate thermal conductivity. Such microgears will not display an homogeneous temperature distribution, but a central temperature higher than that at the gear edges. The temperature at each summit will be lower than the temperature at the clefts. As a result, the temperature gradient of the solvent along the gear edge is still present, which is the crucial point for the motion of the hot microgear.

When coupling the microgear to an external device, net work could be extracted from non-isothermal solutions. As an example of other practical applications, the microgear could be used as a stirring device, which could be locally controlled. Furthermore, when keeping the microgear fixed, the reaction of the thermophoretic force on the tooth edges can result in the motion of surrounding fluid.[78] This can be employed to construct a thermophoretic pump, whose fluid motion is perpendicular to the applied temperature gradient.[79]

4. Thermophoretically Induced Flow Field

The reaction of the thermophoretic force exerted by the non-isothermal solvent on a colloidal particle will in turn result into the solvent motion. To characterize the fluid motion, it is standard to solve the Stokes equation[67] which applies in the low Reynolds regime. The analytical expression of the velocity field around a spherical particle in the presence of a temperature gradient can be obtained by implicitly assuming that the boundary layer approximation is valid (short-range particle-solvent interactions), and the case of an incompressible liquid. The cases of a fixed and a moving colloid in a temperature gradient are important to be analyzed separately.

In the case of a fixed colloid in a temperature gradient the normal component of the flow field at the particle surface vanishes, and the integral of the stress tensor over the particle surface corresponding to the thermophoretic force $\mathbf{f}_T$. The resulting stationary flow field[80,81] is,

$$\mathbf{v}(\mathbf{r}) = -\frac{1}{8\pi\eta r}(\hat{\mathbf{r}}\hat{\mathbf{r}} + \mathbf{I}) \cdot \mathbf{f}_T + \frac{R^2}{8\pi\eta r^3}(3\hat{\mathbf{r}}\hat{\mathbf{r}} - \mathbf{I}) \cdot \mathbf{f}_T. \tag{10}$$

Here, R is the particle radius, η the solvent dynamic viscosity, $\hat{\mathbf{r}} = \mathbf{r}/|\mathbf{r}|$ and $\mathbf{I}$ the unit tensor. The flow is the superposition of a Stokeslet and a source-dipole. Equation (10) indicates that the flow velocity around a fixed particle in a temperature gradient has opposite direction to the thermophoretic force, and that it is of long range since it decays linearly with the inverse distance from the particle center.

In case the colloid can freely move, the thermophoretic force induces a drift in the colloid. The colloid thermophoretic velocity $\mathbf{u}_T$ results from the balance of the thermophoretic and the friction forces $\mathbf{f}_T = \gamma\mathbf{u}_T$, with γ the friction coefficient, which can be approximated to be spatially constant.[82]

The solution of Stokes equation needs to consider now the particle drift and that the balancing forces on the particle result in a vanishing integral of stress tensor over the particle surface. The obtained velocity flow field[67,83] reads,

$$\mathbf{v}(\mathbf{r}) = \frac{R^3}{2r^3}(3\hat{\mathbf{r}}\hat{\mathbf{r}} - \mathbf{I}) \cdot \mathbf{u}_T. \tag{11}$$

The flow velocity on the axis across the colloidal center and along the temperature gradient has now the same direction as the thermophoretic force, and decays with the inverse of the distance cubed. This is much faster than in the case of a fixed particle. The flow fields induced by a thermophilic particle fixed between walls at different temperatures, and freely moving in a temperature gradient are obtained by simulations and shown in Fig. 4. A very precise quantitative agreement with Eqs. (10) and (11) has been demonstrated.[78]

Note that the flow field induced by a sedimenting colloidal particle (the Rotne–Prager–Yamakawa tensor[84,85]) has similar structure to the one in Eq. (10) of a fixed thermophoretic particle. Both systems are though intrinsically different, sedimentation refers always to a moving particle instead of a fixed one, and in contrast with a phoretic force, gravity is an external force applied directly on the particle. The two velocity fields differ in the prefactors, and more significantly the flow induced by a sedimenting particle has the same direction as the gravitational force, while the flow

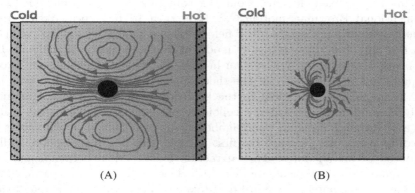

Fig. 4. Cross-section of the flow field induced around a thermophilic colloid in an externally imposed temperature gradient as measured in simulations. (A) Colloid fixed between parallel cold and hot walls. (B) Freely drifting colloid. Small red arrows indicate the flow velocity direction and intensity, while the thick blue lines correspond to the flow stream lines.

induced by a temperature gradient of a fixed particle has opposite sign to the corresponding thermophoretic force.

The importance of hydrodynamic interactions in thermophoretic phenomena has been long a subject of debate.[25,81] The most important example is the explanation of the size independence of thermal diffusion coefficient of a dilute high-molecular mass polymer solution.[83,86-88] From the discussion presented here, it is straightforward to argue that the thermophoretically induced flow field around a fixed colloidal particle accounts for the effect of the thermophoretic force, while the flow around a moving colloid accounts for the combination of the thermophoretic and the friction forces. The long ranged hydrodynamic contribution of the flow field can explain the size independence of thermal diffusion coefficient of a dilute high-weight polymer solution, as a result of the canceling contributions from the size dependence of the thermophoretic force and the self-diffusion coefficient.[18]

The existence of a thermophoretically induced fluid flow has interest not only from the fundamental point of view, but can also find numerous practical applications. Following, we present and discuss two of these applications, the existence of inter-colloidal hydrodynamic attraction, and the possibility of designing a single particle thermophoretic pump.

4.1. *Thermophoretic crystals*

When freely moving colloidal particles are confined between walls at different temperatures, the directional thermophoretic force will naturally drive them towards one of the walls. The colloids may then stay at an averagely fixed distance from the confining wall performing a two-dimensional Brownian motion.[80,89] It is then to be expected that the thermophoretically induced flow field is related to that of a fixed particle in Eq. (10), but asymmetrically modified by the presence of the wall. In contrast to the symmetric system in Fig. 4A where the colloid was fixed equidistant from both walls, the thermophoretically induced flow field of a thermophilic colloid close to the hot wall in Fig. 5 has now a strong lateral component toward the colloidal sphere and parallel to the wall. If a second particle is in the neighborhood of the colloid, it will suffer a hydrodynamic drag toward the first particle, this is a thermophoretically induced attraction force. This attraction force is then perpendicular to the temperature gradient, but will still depend on the same parameters as the thermophoretic force in Eq. (6), and will decrease as function of the distance. When the lateral flow is strong enough (high $|\alpha_T|$), the attraction force can be larger than other repulsive contributions or than thermal fluctuations, originating stable colloid aggregation. Such 2D colloidal thermophoretic crystals, have indeed been experimentally observed,[80,89]

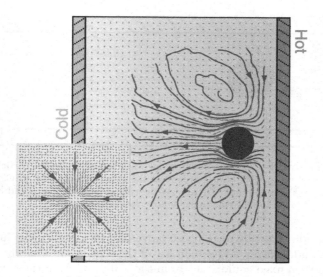

Fig. 5.   Cross-section of the flow field around a thermophilic particle fixed in a near-wall environment. The inset shows the velocity field in a plane parallel to the walls and between the colloid and the hot wall.

and show to form and be stable only in the presence of a temperature gradient. Experiments are performed with thermophobic colloids such that the colloid accumulation takes place in the cold wall, in contrast to the simulation result here displayed. Note that if the colloid would be fixed by an external force to the opposite wall corresponding to their thermophoretic affinity (e.g., thermophobic colloid fixed at the hot wall) the contribution of the thermophoretic flow would induce a repulsive interaction.

4.2. *Thermophoretic pump*

When a colloidal particle is fixed in space, and an alternating temperature gradient (as that in Fig. 6) is switched on, the originally quiescent fluid is initially accelerated by the thermophoretic force. The flow will then arrive to the regions of inverting the temperature gradient with a non-vanishing velocity. This, together with the mass conservation condition, results in a continuous net flow across the whole system,[78] as can be clearly observed in Fig. 6. At the same time, the lateral boundary wall and the colloid surface exerts a frictional force on the flowing fluid, which gradually weakens the fluid acceleration until a constant flow is reached. In the stationary state, the thermophoretic force $\mathbf{f}_T$ is exactly balanced by the friction force.

The simulation results shown in Fig. 6 consider the presence of confining walls in one direction and periodic boundary conditions in the other two.

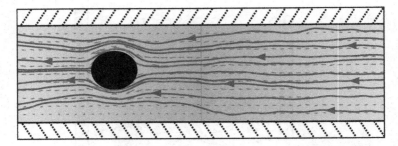

Fig. 6.   Single-particle thermophoretic pump. Thermophilic colloidal particle confined between parallel walls, with both extremes of the tube connected by PBC, and larger temperatures in the tube center. Given that the colloid is thermophilic and has the hot layer on its right, the solvent continuously flows from right to left.

The simulation box is divided in two halves, and temperature is imposed to be hot in a middle layer, and cold in one of the edge layers. One thermophilic colloidal is then fixed equidistant from the walls and from the cold and the hot layer, while no colloidal particle is considered in the neighboring half-box where the temperature gradient has opposite sign. In this configuration the colloid has the hot layer on its right, which induces a flow from right to left. The direction of the flow could be reverted by placing the hot layer on the left of the thermophilic colloid, or by employing a colloid with thermophobic properties instead. The net solvent flow obtained with this particular configuration can be exploited to fabricate a *single-particle microfluidic pump*. This pump has not yet been experimentally realized, but can be built in various ways, for example by fixing one or various colloids in a sequence of temperature gradients with alternating signs. The implementation of this pump does not require the presence of any movable part. The direction of the flow is determined by the orientation of the alternating temperature gradients and the thermophoretic properties of the employed particle.

The so-called Knudsen pumps[90,91] are also pumps operated without any moving part and with a temperature gradient along the walls of a micro-channel. The driving mechanism is though completely different from the one presented here. The Knudsen pumps are driven by *thermal creep* in rarefied gas,[92,93] while here we discuss a pump driven by liquid thermophoretic forces. Thermal creep flow occurs when the molecules of a rarefied gas interact with walls that have a position dependent temperature. The molecules in high temperature region can transfer more momentum to the wall than those in low temperature region, such that the gas exerts a net force on the wall against the temperature gradient.[93] This in turn translates into a flow that goes always from cold to warm areas. In the present thermophoretic pumps, flow can occur in both directions. A different family of microfluidic manipulation that has been widely used[94,95] is based on the existence of a surface tension gradient in the direction of a temperature gradient. For example, liquid droplets or films on a surface can be transported along or against a temperature gradient.[96,97] A very recent simulation work shows that liquids can also be pumped by a symmetric temperature gradient through a composite nanochannel,[98] in which one half of the channel wall has a low fluid-wall surface energy while the other half has a high one. Essentially, the physical mechanism is the same to the pump we present here, since the single particle can be regarded as a curved surface or as building-block of planar walls. The variant discussed here[78] is based on the properties of a single particle which provides an important additional degree of flexibility in the design of microfluidic devices.

5. Conclusions

The recent experimental and theoretical developments in the investigation of thermophoresis pushes forward this old-known phenomena as a promising tool for new-developing applications. Thermophoresis provides an alternative strategy to design synthetic microswimmers, micromotors, or micro-pumps, which have become a promising tool in the field of microfluidics. Temperature gradients can be locally controlled at very small scales, allowing to engineer lab-on-chip devices with effects that could be switched on and off by the local application of temperature sources such as lasers. Furthermore, as discussed in this chapter, all applications of thermophoresis depend not only on the system geometry, the intensity and direction of the applied temperature gradient, but also, and very importantly, on the material properties. This dependence is described by the thermal diffusion factor α_T, and this coefficient is characteristic of each material, or each combination of materials, as well as various other ambient conditions like the average temperature, pressure, concentration, or the solvent pH. Moreover, the competition of thermophoresis with other effects like magnetic properties, or thermoelectricity[99,100] have interestingly shown the existence of materials whose properties vary from thermophobic to thermophilic. In this way, the properties of the thermophoretic machines are very subtly dependent on all the different contributions, providing a very large versatility to these devices, and a fertile investigation field.

On the other hand, some of these devices, like the thermophoretic gear, could also become a very valuable tool to investigate the thermophoretic properties of a wide class of systems. Until now a requirement to determine the thermodiffusion factor, or equivalently the Soret coefficient, has been that the investigated system should be a solution. Therefore, materials that would for example precipitate in solution like gold in water, could be investigated now by means of thermophoretic pumps or microgears.

Acknowledgments

We would like to thank enlightening discussions to Simone Wiegand, Daniel Lüsebrink, Adam Wysocki, Andrea Costanzo, Gerrit Vliegenthart, Roland Winkler, and Gerhard Gompper.

References

1. C. Ludwig, *Sitzungsber. Preuss. Akad. Wiss., Phys. Math. Kl.* **20**, 539 (1856).
2. C. Soret, *Arch. Sci. Phys. Nat.* **3**, 48 (1879).
3. K. Clusius and G. Dickel, *Naturwissenschaften* **26**, 546–546 (1938).
4. R. C. Jones and F. H. Furry, *Rev. Mod. Phys.* **18**, 151 (1946).
5. K. Ghorayeb, A. Firoozabadi and T. Anraku, *SPE Journal* **18**, 114–123 (2003).
6. J. Giddings, *Science* **260**, 1456–1465 (1993).
7. S. Wiegand, *J. Phys.: Condens. Matter* **16**, R357–R379 (2004).
8. R. Piazza and A. Parola, *J. Phys.: Condens. Matter* **20**, 153102 (2008).
9. S. Duhr and D. Braun, *Proc. Natl. Acad. Sci.* **103**, 19678–19682 (2006).
10. H. R. Jiang, H. Wada, N. Yoshinaga and M. Sano, *Phys. Rev. Lett.* **102**, 208301 (2009).
11. C. J. Wienken, P. Baaske, U. Rothbauer, D. Braun and S. Duhr, *Nat. Commun.* **1**, 100 (2010).
12. P. Baaske, F. M. Weinert, S. Duhr, K. H. Lemke, M. J. Russell and D. Braun, *Proc. Natl. Acad. Sci.* **104**, 9346 (2007).
13. I. Budin, R. J. Bruckner and J. W. Szostak, *J. Am. Chem. Soc.* **131**, 9628 (2009).
14. S. R. de Groot and P. Mazur, *Nonequilibrium thermodynamics* (Dover, New York, 1984).
15. N. van Kampen, *J. Phys. Chem. Solids* **49**, 673 (1988).
16. M. Yang and M. Ripoll, *Phys. Rev. E* **87**, 062110 (2013).
17. M. E. Widder and U. M. Titulaer, *Physica A* **154**, 452 (1989).
18. M. Yang and M. Ripoll, *J. Phys.: Condens. Matter* **24**, 195101 (2012).
19. R. Piazza and A. Guarino, *Phys. Rev. Lett.* **88**, 208302 (2002).
20. S. Duhr and D. Braun, *Phys. Rev. Lett.* **96**, 168301 (2006).
21. S. A. Putnam, D. G. Cahill and G. C. L. Wong, *Langmuir* **23**, 9228 (2007).
22. M. Braibanti, D. Vigolo and R. Piazza, *Phys. Rev. Lett.* **100**, 108303 (2008).
23. E. Bringuier and A. Bourdon, *Phys. Rev. E* **67**, 011404 (2003).
24. E. Bringuier and A. Bourdon, *J. Non-Equil. Thermodyn.* **32**, 221–229 (2007).

25. J. K. G. Dhont, S. Wiegand, S. Duhr and D. Braun, *Langmuir* **23**, 1674 (2007).
26. J. K. G. Dhont and W. J. Briels, *Eur. Phys. J. E* **25**, 61 (2008).
27. A. Würger, *Langmuir* **25**, 6696 (2009).
28. G. Galliéro and S. Volz, *J. Chem. Phys.* **128**, 064505 (2008).
29. D. Lüsebrink, M. Yang and M. Ripoll, *J. Phys.: Condens. Matter* **24**, 284132 (2012).
30. A. Würger, *Rep. Prog. Phys.* **73**, 126601 (2010).
31. J. K. G. Dhont, *J. Chem. Phys.* **120**, 1632–1641 (2004).
32. J. K. G. Dhont, *J. Chem. Phys.* **120**, 1642–1653 (2004).
33. H. Ning, J. Buitenhuis, J. K. G. Dhont and S. Wiegand, *J. Chem. Phys.* **125**, 204911 (2006).
34. B. Hafskjold, T. Ikeshoji and S. K. Ratkje, *Molecular Phys.* **80**, 1389–1412 (1993).
35. P. A. Artola and B. Rousseau, *Phys. Rev. Lett.* **98**, 125901 (2007).
36. D. Reith and F. Müller-Plathe, *J. Chem. Phys.* **112**, 2436–2443 (2000).
37. C. Debuschewitz and W. Köhler, *Phys. Rev. Lett.* **87**, 055901 (2001).
38. G. Galliéro, B. Duguay, J.-P. Caltagirone and F. Montel, *Fluid Phase Equilibria* **208**, 171–188 (2003).
39. F. Zheng, *Adv. Colloid Interface Sci.* **97**, 253–276 (2002).
40. C. Nieto-Draghi, J. Bonet Avalos and B. Rousseau, *J. Chem. Phys.* **122**, 114503 (2005).
41. F. Römer, F. Bresme, J. Muscatello, D. Bedeaux and J. M. Rubí, *Phys. Rev. Lett.* **108**, 105901 (2012).
42. J. Luettmer-Strathmann, *J. Chem. Phys.* **119**, 2892–2902 (2003).
43. B. Rousseau, C. Nieto-Draghi and J. Bonet Avalos, *Europhys. Lett.* **67**, 976–982 (2004).
44. M. Vladkov and J. L. Barrat, *Nano Lett.* **6**, 1224–1228 (2006).
45. L. Joly, S. Merabia and J. L. Barrat, *Europhys. Lett.* **94**, 50007 (2011).
46. G. R. McNamara and G. Zanetti, *Phys. Rev. Lett.* **61**, 2332 (1988).
47. P. J. Hoogerbrugge and J. M. V. A. Koelman, *Europhys. Lett.* **19**, 155–160 (1992).
48. F. J. Alexander, S. Chen and J. D. Sterling, *Phys. Rev. E* **47**, 2249(R) (1993).
49. P. Lallemand and L.-S. Luo, *Phys. Rev. E* **68**, 036706 (2003).
50. P. Español, *Europhys. Lett.* **40**, 631 (1997).
51. J. Bonet-Avalós and A. D. Mackie, *Europhys. Lett.* **40**, 141 (1997).
52. M. Ripoll and M. H. Ernst, *Phys. Rev. E* **71**, 041104 (2005).
53. A. Malevanets and R. Kapral, *J. Chem. Phys.* **110**, 8605–8613 (1999).
54. T. Ihle and D. M. Kroll, *Phys. Rev. E* **63**, 020201(R) (2001).
55. J. T. Padding and A. A. Louis, *Phys. Rev. E* **74**, 031402 (2006).
56. A. Malevanets and R. Kapral, *J. Chem. Phys.* **112**, 7260–7269 (2000).
57. M. Ripoll, K. Mussawisade, R. G. Winkler and G. Gompper, *EPL* **68**, 106–112 (2004).
58. R. Kapral, *Adv. Chem. Phys.* **140**, 89–146 (2008).

59. G. Gompper, T. Ihle, D. M. Kroll and R. G. Winkler, *Adv. Polym. Sci.* **221**, 1–87 (2009).
60. D. Lüsebrink and M. Ripoll, *J. Chem. Phys.* **136**, 084106 (2012).
61. D. Lüsebrink and M. Ripoll, *J. Chem. Phys.* **137**, 194904 (2012).
62. J. Howard, *Mechanics of Motor Proteins and the Cytoskeleton* (Sinauer, New York, 2000).
63. H. C. Berg, *E. coli in Motion* (Springer, New York, 2003).
64. F. Jülicher, A. Ajdari and J. Prost, *Rev. Mod. Phys.* **69**, 1269 (1997).
65. W. F. Paxton, K. C. Kistler, C. C. Olmeda, A. Sen, S. K. S. Angelo, Y. Cao, T. E. Mallouk, P. E. Lammert and V. H. Crespi, *J. Am. Chem. Soc.* **126**, 13424 (2004).
66. R. Dreyfus, J. Baudry, M. L. Roper, M. Fermigier, H. A. Stone and J. Bibette, *Nature* **437**, 862 (2005).
67. J. L. Anderson, *Annu. Rev. Fluid Mech.* **21**, 61 (1989).
68. R. Golestanian, T. B. Liverpool and A. Ajdari, *Phys. Rev. Lett.* **94**, 220801 (2005).
69. G. Rückner and R. Kapral, *Phys. Rev. Lett.* **98**, 150603 (2007).
70. P. de Buyl and R. Kapral, *Nanoscale* **5**, 1337 (2013).
71. J. R. Howse, R. A. L. Jones, A. J. Ryan, T. Gough, R. Vafabakhsh and R. Golestanian, *Phys. Rev. Lett.* **99**, 048102 (2007).
72. L. F. Valadares, Y.-G. Tao, N. S. Zacharia, V. Kitaev, F. Galembeck, R. Kapral and G. A. Ozin, *Small* **6**, 565 (2010).
73. H. R. Jiang, N. Yoshinaga and M. Sano, *Phys. Rev. Lett.* **105**, 268302 (2010).
74. M. Yang and M. Ripoll, *Phys. Rev. E* **84**, 061401 (2011).
75. M. Yang, A. Wysocki and M. Ripoll, *Soft Matter*, (2014), DOI: 10.1039/c4sm00621f.
76. M. Yang and M. Ripoll, *Soft Matter* **10**, 1006 (2014).
77. P. G. Saffman and M. Delbrück, *Proc. Natl. Acad. Sci.* **72**, 3111 (1975).
78. M. Yang and M. Ripoll, *Soft Matter* **9**, 4661 (2013).
79. M. Yang and M. Ripoll (2014). Thermophoretic ratcheted micropumps, (preprint).
80. R. Di Leonardo, F. Ianni and G. Ruocco, *Langmuir* **25**, 4247 (2009).
81. J. Morthomas and A. Würger, *Phys. Rev. E* **81**, 051405 (2010).
82. M. Yang and M. Ripoll, *J. Chem. Phys.* **136**, 204508 (2012).
83. A. Würger, *Phys. Rev. Lett.* **98**, 138301 (2007).
84. J. Rotne and S. Prager, *J. Chem. Phys.* **50**, 4831 (1969).
85. H. Yamakawa, *J. Chem. Phys.* **53**, 436 (1970).
86. F. Brochard and P. G. de Gennes, *C. R. Acad. Sci. Paris II* **293**, 1025 (1981).
87. J. Rauch and W. Köhler, *Macromolecules* **38**, 3571–3573 (2005).
88. E. Bringuier, *Physica A* **389**, 4545–4551 (2010).
89. F. M. Weinert and D. Braun, *Phys. Rev. Lett.* **101**, 168301 (2008).
90. M. Knudsen, *Ann. Phys. (Leipzig)* **336**, 205 (1909).
91. S. McNamara and Y. B. Gianchandani, *J. Microelectromech. Syst.* **14**, 741 (2005).
92. J. C. Maxwell, *Philos. Trans. R. Soc. Lond.* **170**, 231 (1879).

93. Y. Sone, *Annu. Rev. Fluid Mech.* **32**, 779 (2000).
94. H. A. Stone, A. D. Stroock and A. Ajdari, *Annu. Rev. Fluid Mech.* **36**, 381 (2004).
95. T. M. Squires and S. R. Quake, *Rev. Mod. Phys.* **77**, 977 (2005).
96. A. M. Cazabat, F. Heslot, S. M. Troian and P. Carles, *Nature* **346**, 824 (1990).
97. J. B. Brzoska, F. Brochard-Wyart and F. Rondelez, *Langmuir* **9**, 2220 (1993).
98. C. Liu and Z. Li, *Phys. Rev. Lett.* **105**, 174501 (2010).
99. D. Vigolo, S. Buzzaccaro and R. Piazza, *Langmuir* **26**, 7792 (2010).
100. A. Würger, *Phys. Rev. Lett.* **101**, 108302 (2008).

Chapter 8

HYDRODYNAMICS MEDIATED COLLECTIVE MOTIONS IN POPULATIONS OF MICRODROPLETS

Shashi Thutupalli[*,†], Jean-Baptiste Fleury[‡], Ulf D. Schiller[§], Gerhard Gompper[§], Stephan Herminghaus[†] and Ralf Seemann[†,‡]

Department of Physics, Princeton University, Princeton, USA

†*Department of Dynamics of Complex Fluids, Max Planck Institute for Dynamics and Self-Organisation, Göttingen, Germany*

‡*Experimental Physics, Saarland University, 66123 Saarbrücken, Germany*

§*Theoretical Soft Matter and Biophysics, Institute of Complex Systems, Forschungszentrum Jülich, 52425 Jülich, Germany*

1. Introduction

The collective dynamics in a self-organizing many-body system presents us with innumerable challenges in understanding how the microscopic interactions between the individual units of a system can give rise to its overall complex, emergent behavior. Vortices in superconductors, bursts of electrical activity in neural networks, self-organized structures during sedimentation of particles in a fluid, flocking of birds or fish etc. are all fascinating manifestations of such non-equilibrium self-organizing behavior.[1] One of the fundamental limitations in understanding such systems lies in the lack of general principles governing non-equilibrium many-body dynamics. This situation is often exacerbated by the nonlinear microscopic interactions among the individual units of the system, which makes any analytical treatment untractable.

In view of these challenges, recently systems of microfluidic droplets have been increasingly used as a table-top system to make inroads into the understanding of out-of-equilibrium many-body collective dynamics.[2-7] In these experiments, microdroplets are driven far from equilibrium[8] by imposing a flow using a microfluidic device and the motions of the individual droplets are correlated due to a long range coupling mediated by hydrodynamic flow fields. Such interactions result in large-scale collective motions of the droplets such as longitudinal and transverse phonon-like motion and Burgers shock waves.

A similar out-of-equilibrium system is a population of hydrodynamically interacting self-propelled microdroplets.[9-11] The self-propulsion of the droplets generates a hydrodynamic flow field in the external fluid, due to which there are interactions between the droplet motions. Such interactions between the swimming droplets can lead to the formation of swarms in the droplet populations.[9,12] Such realizations of non-equilibrium systems, termed as active matter,[13] are an intensely researched topic in recent times, mainly in the context of non-equilibrium statistical mechanics and biology.

A fundamental difference between the self-propelled droplets and the flowing microdroplets is that the source of energy for the self-propulsion resides on each individual droplet while the flow for the collective phonon modes that keeps the system away from equilibrium is imposed externally. However, the linear regime of Stokes flows and interactions due to hydrodynamics are similar features of both and therefore one expects that there will be common signatures of the collective dynamics in both systems. In particular, under appropriate circumstances, both systems (when studied in two dimensions) provide a well-characterizable many-body problem with long-range interactions between the individual units (defined by the characteristic algebraic decay in their hydrodynamic flow fields). In 2D, the dynamics of each individual unit (droplet) in both the systems can be tracked perfectly and hence quantitative comparison between theory and experiment is possible.

Here we present experiments of both the systems mentioned above, i.e., self-propelled microdroplets and microdroplets flowing in a microfluidic channel. In particular, we focus on the collective dynamics of the droplet populations that show up due to hydrodynamics mediated coupling between the droplet motions. In the first part, we present results about the collective behavior of self-propelling liquid droplets which closely mimic the locomotion of some protozoal organisms, so-called squirmers. For the sake of simplicity, we concentrate on quasi-two-dimensional settings, although our swimmers provide a fully three-dimensional propulsion scheme. We find strong polar correlation of the locomotion velocities of neighboring droplets, leading to the formation of ordered rafts or swarms of droplets. In the second part, we discuss long-living coupled transverse and longitudinal oscillations in dense and regular arrangements of flowing, flattened microfluidic droplets. These collective oscillations are driven by hydrodynamic interactions between the confined droplets and can be excited in a controlled way. The observed transverse modes are acoustic phonons and can be described by a linearized far-field theory. The longitudinal modes arise from a nonlinear mode coupling due to the lateral variation of the confined flow field.

2. Experimental Techniques

Microfluidic devices were fabricated from PDMS (Sylgard 184, Dow Corning) using standard soft lithographic protocols and bonded to glass slides. The flow rates were volume-controlled using syringe pumps. Mono-disperse water-in-oil droplets are generated with suitable aqueous and oil phases containing surfactants using step-emulsification.[14]

For the self-propelled droplets, the aqueous phase includes a source of bromine either in the form of dilute (25–50 mM) bromine water or a modified mixture of the Belousov–Zhabotinsky (BZ) reaction mixture. The non-oscillating BZ reaction mixture consists of two parts: (i) 50 mM sulphuric acid (H_2SO_4) and 28 mM sodium bromate ($NaBrO_3$) (ii) 400 mM malonic acid ($C_3H_4O_4$) and 2.7 mM ferroin ($C_{36}H_{24}FeN_6O_4S$). The two parts are mixed just prior to droplet formation. Mono-olein (*rac*-Glycerol-1-Mono-oleate, SigmaAldrich) is used as a surfactant with concentration of (12.5–500) mM, which is much higher than its critical micellar concentration (CMC ≈ 1.5 mM) in the oil phase (squalane, SigmaAldrich). Droplets of the aqueous phase are produced either by hand or using a microfluidic device as shown in Fig. 1 and led into a hydrophobised glass capillary or sandwiched between two hydrophobic glass slides separated by a PDMS spacer of $\sim 100\,\mu$m thickness. The glass slides (and capillaries) are cleaned using isopropyl alcohol, dried and plasma cleaned for 90 seconds. A drop of commercially available Nano-Protect[a] is then used to coat the glass surfaces which are then heated in an oven at 65° for 30 minutes to render them hydrophobic.

For the passive droplets, the height of the PDMS micro-channel is approximately 120 μm and the typical channel width W is about 210 μm.

---

[a]Nano-protect is a commercial hydrophobizing agent made by W5 carcare to make car windshields hydrphobic.

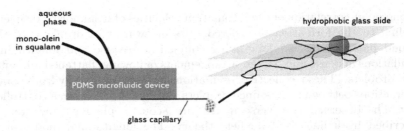

Fig. 1. Left: Microfluidic production of squirmer droplets. The droplet production is by a step emulsification droplet production unit. Right: The squirmers are observed in quasi-one- and two-dimensional environments by confining glass slides or capillaries. The black lines represent the tracks of the swimmers.

Due to the thickness of the device, the aqueous droplets (Millipore™) are flattened. To facilitate the formation of monodisperse water droplets in oil and to avoid undesired droplet coalescence, the non-ionic surfactant Sorbitan Mono-oleate (Span80, conc. 2%wt, Sigma Aldrich) is dissolved in the oily phase (hexadecane, Fluka). The typical flow velocity of the continuous oil phase is $u_{\text{oil}} \approx 500\,\mu\text{m/s}$ resulting in droplet velocities of $u_d \approx 250\,\mu\text{m/s}$. The corresponding Reynolds and Peclet number are Re $= \rho u_{\text{oil}} R/\eta \approx 10^{-2}$ and Pe $= u_{\text{oil}} R/D \approx 10^8$, where R is the droplet radius and $D \approx 10^{-12}\text{cm}^2/\text{s}$ is the diffusion coefficient of a droplet calculated with the formulas given in Ref. 5.

Droplet motion is recorded by a CCD camera (PCO 1200) or a high speed CMOS camera (Phantom Miro, Vision Research) at frames rates between 0.5 Hz and 1000 Hz. Trajectories are obtained by tracking the droplet positions using MATLAB (MathWorks) and Image-Pro Plus (MediaCybernetics). The flow fields around the squirmer are recorded by Micro Particle Image Velocimetry (μPIV).[15] For μPIV experiments the oil phase is seeded with fluorescent tracer beads (200 nm green fluorescent polystyrene beads, Duke Scientific) and fluorescence microscopy is used to record double images of the tracer particles. The flow fields are determined from the double images using the open-source code PIVlab.[16]

3. Self-Propelled Squirming Droplets

We present a simple model squirmer consisting of an aqueous droplet moving in an oil "background" phase. Propulsion arises due to the spontaneous bromination of mono-olein (*rac*-glycerol-1-mono-oleate) as the surfactant. The latter is abundantly present in the oil phase, such that the droplet interface is covered by a dense surfactant monolayer. The bromine "fuel" is supplied from inside the droplet, such that bromination proceeds mainly at the droplet surface. It results in saturation of the C=C double bond in the alkyl chain of the surfactant, thereby rendering it a weaker surfactant. As we will see below, this results in a self-sustained bromination gradient along the drop surface, which propels the droplet due to Marangoni stresses.

We demonstrate this squirmer scheme with nanoliter droplets containing 25 mM/l Bromine water in a continuous oil phase of squalane containing 50 mM/l mono-olein (MO). The critical micelle concentration (CMC) is 1.5 mM/l. Although we here confine the droplets to a space of reduced dimension, it should be kept in mind that their propulsion mechanism is inherently three-dimensional, thereby closely mimicking the behavior of protozoal organisms. This clearly distinguishes our system from "striders" which are inherently bound to an interface.[17–21] Figure 2B shows the trajectory of a single squirmer droplet over a duration of 400 seconds. The velocity of the droplet is about 15 μm/sec. The trajectory is reminiscent of a random walk, with a persistence length which is larger than the droplet size and clearly far beyond what would be expected for Brownian motion. A particularly important observation is that the trajectory crosses itself. This is in sharp contrast to other schemes, where the propulsion mechanism itself changes the surrounding medium strongly enough to prevent self-crossing of the path.[17] That nothing like this happens here is demonstrated even more convincingly in Fig. 2D, which shows a time lapse representation of six droplets moving in a micro-channel. As two drops touch each other, they reverse their direction of motion and perambulate the channel again, without significant reduction in velocity. This can be attributed to the abundance of surfactant in the oil phase, which ensures that the medium is only minutely changed by the "exhaust" of the squirmers. This property makes our system particularly well suited for the study of collective behavior.

3.1. *Mechanism of locomotion*

In order to gain some insight into the propulsion mechanism, let us consider a spherical droplet with radius R. The total coverage, c, of the droplet surface with the MO, either brominated or not, is assumed to be roughly constant and in equilibrium with the micellar phase in the oil. The brominated fractional coverage shall be called b. If the droplet moves, there is (in the rest frame of the droplet) an axisymmetric flow field $u(\theta)$ along its surface. The equation of motion for b is

$$\frac{\partial b}{\partial t} = k(b_0 - b) + \text{div}(D_i \, \text{grad} \, b - ub) \tag{1}$$

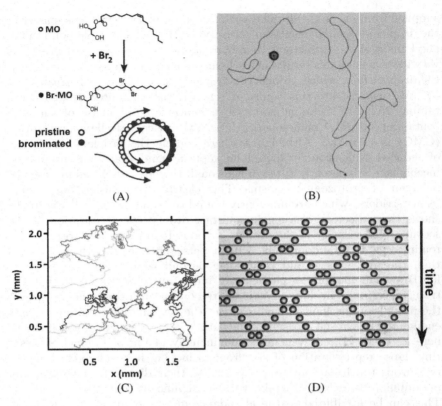

(A)   (B)

(C)   (D)

Fig. 2.   (A) Schematic of a microdroplet squirmer. Bromination of the pristine surfactant molecule increases the tension of the droplet surface from 1.3 mN/m to 2.7 mN/m. The convective flow pattern (shown in the rest frame of the droplet) is accompanied by a gradient in the bromination density. The corresponding Marangoni stress propels the droplet. (B) Path of a single squirmer droplet. The persistence length is clearly large compared to the droplet radius, indicating propelled motion (Scale bar: 300 microns). (C) Droplet trajectories over an interval of ~90 s. (D) Time lapse series of droplets in a micro-channel whose width is ~100 microns. When two droplets collide, they change direction without noticeable reduction in velocity.

where b_0 is the equilibrium coverage with brominated mono-olein (brMO). It is determined by the bromine supply from inside the droplet and the rate constant, k, of escape of brMO into the oil phase. The first term on the r.h.s. of Eq. (1) describes the balance between bromination and brMO escape. The second term describes the change in the bromination density of the surfactant layer due to transport along the droplet surface. D_i is the diffusivity of the surfactant within the interface.

The droplet motion and the surface flow $u(\theta)$ are accompanied by a flow pattern within the droplet as well as in the neighboring oil, which can be calculated analytically once $u(\theta)$ is known.[22,23] The corresponding viscous tangential stress exerted on the drop surface must be balanced by the Marangoni stress, $\operatorname{grad}\gamma(\theta) = M\operatorname{grad}b(\theta)$, where γ is the surface tension of the surfactant-laden oil/water interface, and $M = d\gamma/db$ is the Marangoni coefficient of the system. Expanding the bromination density in spherical harmonics,

$$b(\theta) = \sum_{m=0}^{\infty} b_m P_m(\cos\theta), \qquad (2)$$

we can express the velocity field[22,23] at the interface as

$$u(\theta) = \frac{M}{\mu\sin\theta} \sum_{m=1}^{\infty} \frac{m(m+1)b_m C_{m+1}^{-1/2}(\cos\theta)}{2m+1}, \qquad (3)$$

where C_n^α denote Gegenbauer polynomials, and μ is a prefactor containing the liquid viscosities outside and inside the droplet. Inserting this into Eq. (1) and exploiting the orthogonality relations of Gegenbauer and Legendre polynomials, we obtain

$$\frac{db_m}{dt} = \left[m(m+1)\left(\frac{b_0 M}{(2m+1)R\mu} - \frac{D_i}{R^2} \right) - k \right] b_m \qquad (4)$$

for all $m > 0$. The different modes decouple, as far as linear stability is concerned. As long as $b_0 M$ is small enough, the expression in brackets is negative, and the resting state is stable against fluctuations. However, when $b_0 M$ exceeds a critical value, the resting state is unstable, and the droplet spontaneously starts to move. It is straightforward to see that for $k < 3D_i/R^2$, this happens first for the lowest mode at $m = 1$. In order to determine the flow profile around a squirming drop, we performed particle image velocimetry using a standard setup (ILA GmbH, Germany). The result as displayed in Fig. 3C shows that the flow profile indeed resembles a field as expected for the lowest order mode, $m = 1$.[24] This corresponds also to the characteristic $1/r^2$ decay of the flow field in a quasi-2D environment.

3.2. *Properties of squirming droplets*

Before we turn to the collective behavior of the squirmer droplets introduced above, we have to dwell some more on their properties. To discuss the steady state droplet velocity V we first reconsider the total surfactant coverage c. This adjusts itself so as to balance the molecular adsorption energy at the water/oil interface and the mutual repulsion of the adsorbed surfactant

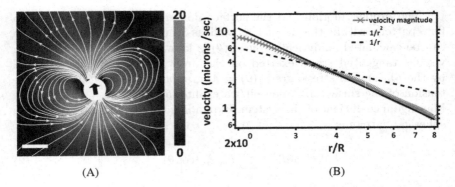

Fig. 3.   (A) Velocity field around a droplet squirmer. The magnitude of the flow velocity (color code) and streamlines along a horizontal section through the center of a squirmer droplet are shown. Scale bar: 100 microns; velocity scale (right) in microns per second. (B) The decay of the velocity field around the droplet follows a power law scaling as $1/r^2$ which causes a long-range interaction between the droplets in a quasi-2D confinement.

molecules. The equilibrium coverage thus represents a minimum in the interfacial energy. As a consequence, the interfacial tension will not change to first order if c is varied. Small deviations of c from its equilibrium value, which come about necessarily for any finite V, will thus be replenished from the surrounding oil phase without substantial Marangoni stresses. For a flow pattern as the one shown in Fig. 3 we approximately have $\operatorname{div} u \propto \cos\theta$. The density of surfactant in the oil phase thus takes the form $\rho \approx \rho_0 + \delta\rho f(r)\cos\theta$, where $\delta\rho \propto V$. $f(r)$ is some function of the radius. If D_m is the diffusivity of the micelles in the oil, the diffusion current, $D_m \partial\rho/\partial r$, must balance the depletion rate at the drop interface, $c \operatorname{div} u = \frac{3Vc}{2R}\cos\theta$.[24] As long as this holds, the only source of appreciable Marangoni stresses is the gradient in bromination density, $b(\theta)$. The drop thus keeps taking up speed according to Eq. (4). This comes to an end when $\delta\rho \approx \rho_0$. The surfactant layer at the leading end of the drop surface can then not be replenished anymore, and c is reduced substantially below its equilibrium value. This leads to an increase of surface tension accompanied by a "backward" Marangoni stress, and thus eventually to a saturation of the velocity.

According to the reasoning above, we expect that $V \approx 2\rho_0 D_m/3c$. There is no literature value for the diffusivity of MO micelles in squalane, but we can estimate it on the basis of the Stokes–Einstein relation assuming the radius of the micelles to be similar to the length of a MO molecule (2.3 nm). Using 36 mPas for the viscosity of the squalane, we obtain $D_m = 2.6 \times 10^{-12}\,\mathrm{m^2/s}$. We thus predict $V/\rho_0 \approx 0.27\,\frac{\mu\mathrm{m/sec}}{\mathrm{mM/l}}$ from the simple consideration above.

In order to measure $V(\rho_0)$, we used a reaction scheme similar to the Belousov–Zhabotinski (BZ) reaction, with reactant concentrations adjusted such as to prevent chemical oscillations to occur (50 mM sulphuric acid, 28 mM sodium bromate, 400 mM malonic acid, and 2.7 mM ferroin).[25,26] This results in a rather constant bromine release rate in the aqueous phase for an extended period of time. Figure 4A shows the dependence of the droplet velocity on the mono-olein concentration in the oil phase. The initial linear increase is in quantitative agreement with the above prediction (dashed line). As the surfactant density is further increased, the complex exchange processes between the water/oil interface and the micelles will finally become the rate limiting step. As a consequence, the velocity is expected to level off, which is again in accordance with our data.

It is thereby clear that the bromination profile across the droplet surface $b(\theta)$ will not be anymore determined by the fastest growing mode as suggested by Eq. (4). Instead, it will acquire a complex shape accounting for the nonlinearity of Eq. (1). It is so far entirely unclear how large the nonlinear contributions to $b(\theta)$ will be. As a consequence, a quantitative estimate of the driving force $b_0 M$ on the basis of the steady state velocities might easily be orders of magnitude off reality. It will be left to forthcoming studies to quantify the driving force and its relation to the other quantities characterizing our squirmer.

Next, we come to the discussion of directionality, which together with the speed characterises the squirmer velocity. For a given swimmer, we calculate its directionality, defined as $\langle \cos \phi \rangle$, where ϕ is the turn angle i.e., the angle between the velocity vectors of the swimmer at equidistant time steps, i.e., every second, as a function of time. As seen in Fig. 4B, which shows the data for surfactant concentration of 100 mM, the directionality increases linearly with time and after ~200 seconds, it reaches a plateau around ~0.4. As seen in Fig. 4C, this remains roughly constant for a range of surfactant concentrations. Therefore, the change in the directionality of the droplet is likely to be a consequence of the driving from within, namely the bromine source. Initially, when there is a surplus of bromine, the droplets get "kicked" around due to the rapidly changing interface conditions. However, as this rate reduces, a balance is reached between the reaction from inside and the replenishment of the surfactant from the outside. At this stage, the surface coverage is maintained around an equilibrium value, thus making the motion homogeneous and directionally persistent. It should be noted, however that this interpretation is preliminary since effects due to dissolution of MO into the aqueous phase have been so far neglected.

During an experiment, the concentration of brMO in the oil phase gradually builds up from the "exhaust" of the many squirmers in a poorly controllable and spatially inhomogeneous fashion. Since the majority of the

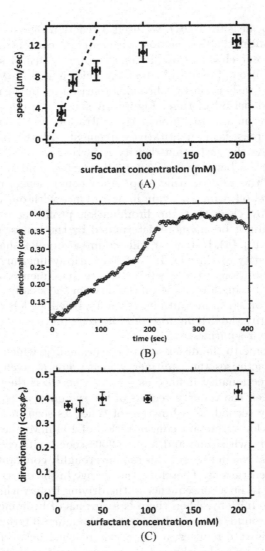

Fig. 4.   Characterization of squirmer velocity. (A) Speed in units of microns per second, as a function of surfactant concentration, where each point is an average of 50 different squirmer droplets, each of diameter 80 microns. The dashed line corresponds to the theoretical prediction (see text). (B) The time varying directionality is plotted for a surfactant concentration of 100 mM. A linear increase is followed by a saturation of the directionality at ∼0.4. (C) The time averaged directionality as a function of surfactant concentration, where each point is an average of ∼50 different squirmer droplets each of diameter ∼80 microns.

brMO becomes trapped in the MO micelles, its effect is mainly to reduce the effective concentration of MO in the oil. Figure 4 tells us that by choosing a high surfactant concentration, where there is almost no sensitivity of the squirmer velocity on the surfactant density, we can also minimize the sensitivity of the velocity to the concentration of brMO, thereby minimizing any unwanted (global) cross talk between the squirmers. This property provides a key feature of our system which makes it particularly suited for experimental studies of collective motion.

3.3. *Collective behavior*

Now we finally turn to the investigation of the collective behavior of our squirming droplets. It is a long standing debate whether physical effects, like hydrodynamic interactions, are *sufficient* to explain textures observed in the swarming behavior of bacteria and other micro-organisms, without having to invoke chemotaxis or other genuinely biological effects.[27–30] Using our model squirmers instead of bacteria, we can tackle this problem from the reverse side, asking for the textures we can observe in dense populations of model squirmers, which are guaranteed to exhibit *no* biological interactions. As we will see, there is indeed considerable structure to unveil even for such simple systems.

We restrict ourselves to effectively two-dimensional systems, not only for the sake of simplicity, but also because most studies so far have concentrated on the two-dimensional case. After generation the droplets were flushed into shallow wells and the flow is stopped. The wells have diameters of a few millimeters and a depth slightly larger than the droplet diameter. Figure 5A shows a typical sample; the arrows indicate the direction of motion of each droplet.

As time proceeds, the formation of rather long-lived clusters of different size is observed.[31] The most striking feature, however, is the significant polar alignment of the velocities of neighboring droplets. In order to quantify this alignment, we use the angular correlation function,

$$C_\vartheta(r) := \langle \delta(r - \mid r_i - r_j \mid) \cos \vartheta_{ij} \rangle_{t,ij} \tag{5}$$

which describes the propensity of velocities of neighboring particles to align with respect to each other. The average runs over all droplet pairs, (ij), as well as over time t. ϑ is the angle between the velocities of droplets i and j and depends on time for each droplet pair.

On the basis of earlier theoretical work, we might under certain circumstances expect a significant polar correlation of the velocities of neighboring droplets,[29,32] just from the hydrodynamic interaction of the squirmers with each other. That this is indeed the case can be seen

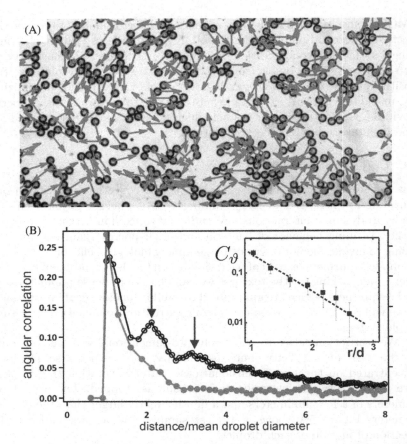

Fig. 5. (A) Top view of a typical experiment. The droplet diameter was 80 microns. The surfactant concentration was 200 mM/l in order to minimize the sensitivity of the droplet motion to the MO and brMO concentrations. The arrows indicate the momentary direction of the droplet motion. (B) The angular correlation of the droplet motion, C_ϑ, as a function of the scaled distance of the droplet centers, r/d. $d = 2R$ is the droplet diameter. The gray curve corresponds to an areal droplet density of 0.46, the black curve to a density of 0.78. The arrows are at multiples of 1.08 d. The inset shows a semilogarithmic plot of the decay of the correlation data. The gray line corresponds to a decay length of 0.6 d, the straight asymptote of the black line represents a decay length of 2.5 d. An oscillation with a period of 1.08 d (decaying over 0.9 d) has been superimposed to fit the data.

in Fig. 5B. The gray curve corresponds to a moderate droplet density of 0.46, which we define as the fraction with respect to the density corresponding to a hexagonal close packing in the plane. We see that there is significant correlation for small distances. More specifically, the angular correlation function decays approximately exponentially away from the contact distance (which is equal to one droplet diameter, $d = 2R$), as can be seen from the inset. The decay constant is about 0.6 droplet diameters (gray line in the inset). The angular correlation thus decays almost completely over one interparticle distance, suggesting that the correlation of the velocities is mediated by the pair interaction of the particles. In fact, it has been predicted theoretically that two adjacent droplets which are propelling themselves by means of low-order spherical harmonic flow fields on their surfaces may attract themselves into a bound state in which they are swimming with virtually parallel velocities.[32,33] In the case of very high Peclet numbers, this can lead to the formation of a bound state in which the particles are swimming with virtual parallel velocities.[32] The attraction becomes weaker with decreasing Peclet number,[33] so that the life time at the bound state decreases. We provide here a first experimental corroboration of this prediction using a purely "physical" system.

There are, however, also pronounced differences with simulation data. Ishikawa and Pedley[29] have performed simulations of the collective behavior of spherical droplets with full hydrodynamic interactions swimming in a monolayer. This system is very similar to ours, but the monolayer in the simulation was freely suspended in the three-dimensional "liquid" such that there was no nearby wall as in our case. This gives rise to long-range hydrodynamic interactions and provides a straightforward explanation why they found velocity correlations ranging up to more than five droplet diameters, in marked contrast to our results.

Correlations become more pronounced as the density of the droplets is increased to an areal density of 0.78. As the black curve in Fig. 5B shows, we observe two significant changes. First, the range of the correlation becomes significantly larger, extending clearly beyond four droplet diameters. Second, we observe the appearance of distinct peaks in the correlation function. The arrows are at multiples of 1.08 droplet diameters, which is close to what one would expect in case of lateral "layering" effects. As the inset shows, the decay length of the polar correlation is now about 2.5 droplet diameters (slope of the black asymptotic line) and has thereby increased by a factor of four. The amplitude of the superposed oscillations (describing the lateral layering) decays over 0.9 droplet diameters. These results suggest the presence of a phase transition occurring at a density somewhere between 0.46 and 0.78, at which a qualitative change in the correlation behaviour takes place.

4. Oscillations of Passive Droplets

We consider passive droplets which are driven by the surrounding continuous phase. The droplets are generated in devices as described in

S. Thutupalli et al.

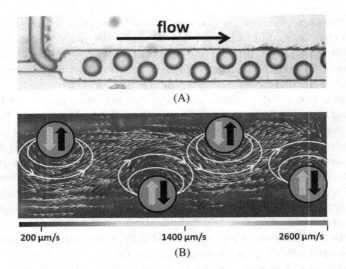

(A)

(B)

Fig. 6. (A) Generation of a microfluidic crystal in a stable zigzag arrangement using a step junction. (B) Flow field determined by PIV in the lab frame. The dipolar hydrodynamic interactions between droplets are sketched as white arrows. The transverse forces resulting from the previous and the trailing droplets are shown as gray and black arrows, respectively.

the experimental section using step-emulsification.[14] After generation the droplets are flushed into a straight microfluidic channel with rectangular cross-section, see Fig. 6.

4.1. Quasi-2D hydrodynamic interactions of microfluidic droplets

The considered droplets are confined between two parallel plates and thus move in a quasi-2D geometry, cf. Fig. 7. The flow and the hydrodynamic interactions in this geometry differ qualitatively from the bulk case due to momentum absorption at the confining plates which leads to screening of the far-field. The fluid flow satisfies no-slip boundary conditions on the channel walls, and since the height H of the channel is small compared to the lateral width W, the velocity gradient in the z-direction is much larger than in the planar directions. In the Darcy approximation $\partial_z^2 u \gg \partial_x^2 u$, $\partial_z^2 u \gg \partial_y^2 u$, the solution of the Stokes equation has a quasi-2D Hele–Shaw form[34]

$$\vec{u}(x,y) = \frac{1}{H}\int_{-H/2}^{H/2} dz\, \frac{z^2 - H^2/4}{2\eta}\nabla p(x,y) = -\frac{H^2}{12\eta}\nabla p(x,y). \tag{6}$$

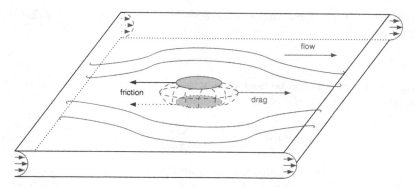

Fig. 7. Schematic illustration of the quasi-2D geometry. When the droplets move relative to the imposed flow, they act as a mass dipole with a sink at their leading edge and a source at their trailing edge. Flattened droplets experience a friction at the top and bottom plates which counteracts the hydrodynamic drag.

At low Reynolds number, the flow is incompressible and Eq. (6) can be written as a Laplace equation $\nabla^2 \phi = 0$ with the potential $\phi = -H^2 p / 12\eta$. A droplet that is moving through the fluid with a velocity $\vec{u}_d$ creates a perturbation of the flow. The moving droplet acts as a momentum monopole whose flux scales as u^2. However, due to the absorption of momentum at the top and bottom plates the flux is not conserved. The absorbed flux scales as u/h, therefore the flow field of the momentum monopole $\partial_r u \propto -u/h$ decays exponentially. Thus, unlike in the bulk case, the leading contribution is the mass dipole created by the droplet which in quasi-2D decays as r^{-2}.[35] Formally, the flow perturbation at a distance $\vec{r}$ from a droplet is obtained by solving the Laplace equation with boundary conditions of zero mass flux (zero perpendicular velocity) through the droplet interface.[36] This gives the dipolar potential

$$\phi_d = -R^2 \delta\vec{u} \cdot \frac{\vec{r}}{r^2}, \tag{7}$$

where $\delta\vec{u} = \vec{u}^\infty - \vec{u}_d$ is the velocity of the droplet relative to the externally imposed flow $\vec{u}^\infty$.

The hydrodynamic drag on an isolated droplet in the potential flow is then

$$\vec{F}_{\text{drag}} = \frac{12\eta}{H^2} \oint \phi d\vec{S} = \frac{24\pi\eta R^2}{H} \left(\frac{1}{2}\vec{u}_d + \delta\vec{u} \right), \tag{8}$$

where the second term in the brackets is due to the self-interaction of the droplet with its dipole.[5]

If the size R of the droplets exceeds the channel height H, they are flattened and experience a friction with the top and bottom plates

$$\vec{F}_{\text{frict}} = -\zeta \vec{u}. \tag{9}$$

Since inertial effects can be neglected, we can use force balance to determine the mobility $\mu = u_d/u^\infty$ of the droplet in an imposed flow

$$\vec{u}_d = \mu \vec{u}^\infty = \left(\frac{1}{2} + \frac{\zeta H}{24\pi\eta R^2} \right)^{-1} \vec{u}^\infty. \tag{10}$$

In an ensemble of droplets, the solution of the Laplace equation is considerably more complicated because the boundary conditions have to be satisfied on all surfaces. Although this is in principle possible using the method of images, the large number of reflections that arise makes it very complicated in practice. One therefore resorts to the leading-order approximation where the drag force is given by a superposition of dipolar flow fields created by the other droplets. This leads to the velocity for the ith droplet in an ensemble[5]

$$\vec{u}_i = \mu \left(\vec{u}^\infty + \sum_{j \neq i} \nabla \phi_d(\vec{r}_i, \vec{r}_j) \right). \tag{11}$$

Under lateral confinement, the inter-droplet potential $\phi_d(\vec{r}_i, \vec{r}_j)$ depends explicitly on the positions of the droplets. Its analytical form is given in Refs. 37 and 5. Although this approximation is based on the assumption that the inter-droplet distance is larger than the droplet size $r_j - r_i \gg R$, we will see below that the predictions based on Eq. (11) work well even for dense droplet trains.

4.2. Controlled excitation of collective oscillations

Having discussed the flow fields and the forces acting on a single droplet and droplets in dense droplet ensembles, we come back to a flowing arrangement of droplets in a straight microfluidic channel. The continuously generated train of droplets initially forms a stable zigzag configuration, where the dipolar forces acting on the neighboring droplets cancel out,[38] see Fig. 6. When the droplets in the initially stable zigzag arrangement are guided around a bend, a defect in the ordering can be achieved, as displayed in Fig. 8. After the bend, the regular zigzag positions are perturbed, the symmetry is broken, and some droplets experience a net hydrodynamic force. The droplets which experience a net force towards the other side of the channel are consequently moved towards the opposite channel wall,

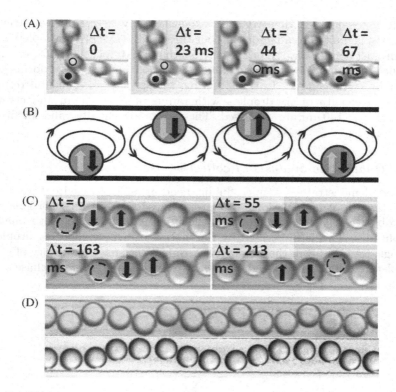

Fig. 8.   (A) Microscopy time series showing the droplet reorganization at a 90° bend. (B) Sketch indicating the dipolar hydrodynamic interactions between the droplets. The transverse forces resulting from leading and trailing droplets in a zigzag configuration with one "mismatch" are shown as gray and black arrows, respectively. (C) Time series of a transverse drop motion for a single "mismatch" as described in (B). Three drops are marked and the net transverse forces are shown as black arrows. (D) Travelling sine waves as generated by periodic "mismatches" using the geometry in (A) having different drop size and different wavelength.

see Fig. 8. Following this transverse movement, a single defect propagates backwards in the co-moving frame of the droplets.

Provided the droplet size relative to the channel width and droplet density are chosen appropriately, defects can be generated periodically. This leads to sine-waves of droplets travelling forward in flow direction, see Fig. 8. The observed collective oscillations are very stable and could be observed for channel lengths up to 10 cm, i.e., after travel distances which are four orders of magnitude larger than a typical droplet radius. The wavelength λ in longitudinal direction depends on both the droplet size and the droplet spacing. Varying these parameters, various wavelengths can be specifically excited, see Fig. 8. However, the experimentally accessible wavelength variation is limited to a factor of about three as the size of the droplets and the droplet density have to be chosen such that the droplets are close enough to affect each other by hydrodynamic interactions, but do not deform each other by steric contact. In the following, the specific experimental results for a wavelength of six droplets with radius of $R = 64\,\mu$m and crystal spacing $a = 140\,\mu$m are discussed. The results are equally valid for other wavelengths.

4.3. *Qualitative analysis of collective oscillations*

As we demonstrated, a microfluidic bend is able to excite transverse collective droplet oscillations. But beside these transverse oscillations, we also observe collective oscillations in the longitudinal direction with smaller amplitudes. This is illustrated by the motion of five neighboring droplets in Figs. 9A and 9B in their co-moving frame. The space trajectory of each droplet describes a figure-eight and each droplet has a constant phase shift

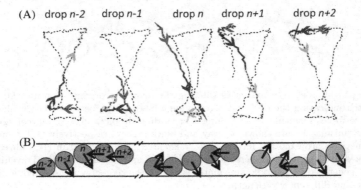

Fig. 9.   (A) Configuration space trajectory of five neighboring droplets in the moving frame of the droplets as extracted from the experiments. To avoid overlaps, the lateral scale of the trajectories was increased. The trajectories of three time intervals are colored in red, blue and green. The flow direction is from left to right. (B) Sketch of the five drops from (A) in a micro-channel with directions of motion in the crystal frame of reference. (C)–(D) Power spectra of the Fourier transform of the droplet oscillations. (C) Transverse modes $y(k,\omega)$ as extracted from experimental results. The white solid curves denote the prediction from the linearized theory for $\omega_\perp(k)$ whereas the white dashed lines correspond to the continuum approximation with nearest neighbor interactions only. (D) Longitudinal modes $x(k,\omega)$. The gray solid curves are the prediction $\omega_\parallel(k)$ for acoustic phonons from the linearized theory whereas the dashed lines correspond to the continuum approximation. The white solid curves represent $2\omega_\perp(k/2)$ and illustrate the coupling of the longitudinal to the transverse modes. The time scale is $\Delta T = R/u_d$.

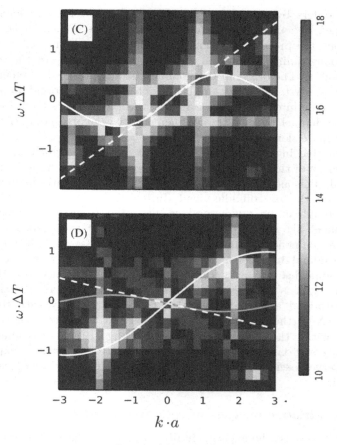

Fig. 9. (*Continued*)

to its neighboring droplets. The figure-eight results from the transverse oscillation with comparable large amplitude $W - 2R \approx 82\,\mu$m given by the lateral confinement, and a longitudinal oscillation with smaller amplitude $a - 2R \approx 12\,\mu$m given by the crystal lattice a, i.e., the droplet density. For large transverse displacement, the droplets closely approach the channel walls, where they slow down as a consequence of the no-slip boundary condition. Thus, the droplets move backward in the co-moving frame when they are at the channel wall and forward when they cross the channel. One full cycle of the longitudinal oscillation is completed while the droplet moves from one channel wall to the other, corresponding to half a transverse cycle. Accordingly, the constant phase shift between the oscillations of neighboring droplets differs by a factor of two between transverse and longitudinal modes, i.e., 55° and 110° for the considered wave. When the transverse oscillations are propagating along the micro-channel, they also trigger a coupled longitudinal excitation due to the nonlinearity of the ensemble flow potential. The origin of the nonlinearities is expected to emerge from two sources: First, the larger amplitude of the transverse oscillations, as the hydrodynamic interactions are a function of the droplets lateral positions across the channel. Second, the confinement, i.e., the presence of the channel wall which are slowing down the droplets.

As plotted in Fig 9A, the figure-eight motion is slightly asymmetric, as the length of the two longitudinal displacements on top and bottom are not equal. This asymmetry results from the initial longitudinal excitation induced by the microfluidic bend during the generation of a packing defect. So it can be understood as an initial trigger of a longitudinal displacement like a *momentum* due to the droplet re-organization at the bend. The influence of this *momentum* can be observed when analyzing the longitudinal displacement which becomes symmetric when the droplets traveled some distance away from the bend. At this distance from the bend, the figure-eight is symmetric within experimental accuracy. Furthermore, the longitudinal amplitude seems to decay very slowly with the travel distance. Nevertheless, it was neither possible to observe the complete damping within the tested channel length of up to 10 cm nor to quantify the relatively small damping within experimental errors. Thus, the experimentally observed oscillations are constant in time at a short distance beyond the bend.

4.4. Quantitative analysis of collective oscillations

In order to discuss the experimentally observed oscillations quantitatively, we extracted the longitudinal and transverse power spectra from the experimental data. The main feature of the obtained power spectra are a few distinct peaks for both modes. These peaks suggest a collective oscillation with a well-defined wavelength propagating in both directions, which was experimentally excited. The observed transverse oscillations can be characterised as acoustic phonon modes by comparing them with the dispersion relation obtained by a linear far-field theory for two-dimensional channel flow:[5,37]

$$\omega_\perp(k) = 2B \sum_{j=1}^{\infty} \sin(jka) \left[1 + \cosh(j\pi\beta)\right]^2 \operatorname{csch}^3(j\pi\beta) \qquad (12)$$

where $B = u_d/u_{\text{oil}}(u_{\text{oil}} - u_d)(\pi^2 R/W^2)\tan(\pi R/W)$, and $\beta = a/W$. Note that $\omega_\perp(k)$ in Eq. (12) has a slightly different form than reported in

Refs. 37 and 5, as it takes into account all linear-order terms in the expansion. The agreement of the analytical dispersion relation with the experimental result is surprising, as the far-field approximation used to derive the analytical dispersion relation generally does stay valid in a system where droplets are almost in contact. One possible reason is that in the relatively strong lateral confinement we study, the higher reflections of the dipoles are so weak that they might still be neglected. Extracting the speed of sound $C \approx \delta w / \delta k_{k \approx 0}$ from the acoustic phonon power spectrum, we obtain the propagation speed $v_p \approx 312 \, \mu\text{m/s}$ of the phonon mode along the droplet crystal in direction of the flow. It can be seen in Fig. 9D that the corresponding longitudinal acoustic dispersion relation does not describe the experimental observation. As already mentioned in the discussion of the figure-eight motion of the droplets, the longitudinal vibrations are coupled to the transverse acoustic modes. By definition, a mode-coupling relation results from an interaction between two vibrational modes. Thus, the longitudinal dispersion relation can be phenomenologically predicted by $\omega_{\parallel}(k) = 2\omega_{\perp}(k/2)$, which is beyond the linear far-field theory.[6] Furthermore, it should be possible to detect the mode-coupling signature by comparing the longitudinal power spectrum with the transverse power spectrum. Close inspection of the experimental features in Figs. 9C and 9D reveals that the position of the twin peaks in the longitudinal spectrum are visible also in the transverse power spectrum at corresponding wave vectors. Vice versa, the main peaks in the transverse spectrum can be observed in the longitudinal spectrum.

Beside the two peaks in each experimental power spectrum which can be described as acoustic transverse phonons and a phenomenological dispersion relation, the power spectra also present weaker peaks. These weaker peaks can be described by a linear dispersion relation having a positive slope for transverse modes and a negative slope for longitudinal modes and result from a slightly varying longitudinal spacing of the droplets. When the droplets move away from their regular crystal positions, they can be regarded as a continuous ensemble, and the finite-differences in the equation of motion are replaced by a continuum approximation. The resulting linear dispersion relations are plotted as dashed lines in Figs. 9C and 9D. The signal of this branch stays weak and constant when the droplets travel downstream. The weak and stable features in the power spectrum supports the experimentally observed stability of the droplet oscillations which did not noticeably change even at distances of 10 cm after the bend. In order to understand the longitudinal and transverse phonons and their non-linear coupling in more detail, we have also performed simulations of droplets in two-dimensional channels,[37] using a mesoscale hydrodynamics simulation technique.[39] This approach requires no approximations, and therefore goes beyond the far-field description. The simulation results to the dispersion relation agree very well with the theoretical prediction.[17] Furthermore, since the Peclet number in the simulations is considerably smaller than in our experiments, the full dispersion relation can be obtained.[37]

5. Conclusions

In the first part of this chapter, we have introduced artificial squirmers which self-propel by a Marangoni flow leading to a dipolar flow field around each droplet. The Marangoni flow is caused by a bromination reaction of surfactant molecules, whereas the chemical reaction has a very weak influence on the environment such that any global chemical coupling of swimmers is excluded. Therefore the model squirmers are ideally suited to study open questions regarding the physical interactions between self-propelling particles and, in particular regarding the role of hydrodynamic interactions on the statistical mechanics of SPP populations. The collective behavior of these squirmers shows strong velocity correlations, which are qualitatively similar to recent simulation results. There are, however, distinct differences which can be attributed in part to simplifications which had to be made in the simulations. Of particular interest is the occurrence of ordered rafts with a lateral correlation extending over more than four droplet diameters, which emerge for large droplet densities.

In the second part of this chapter, we studied passive droplets traveling along a straight microfluidic channel which are driven by the continuous phase. Due to friction with the confining top and bottom wall, the droplets travel slower than the continuous phase resulting in a dipolar flow fields around each droplet. In a certain droplet arrangement, these hydrodynamic interactions lead to coupled phonon modes which can be experimentally excited. The emerging transverse oscillations could be described as acoustic phonons by a linearized far-field theory even in the dense system where droplets are almost in contact. The considerable amplitude of the excited transverse oscillations additionally leads to a nonlinear coupling of longitudinal and transverse modes. This mode-coupling is due to the no-slip boundary condition at the channel wall and the lateral variation of the hydrodynamic interactions across the channel. The coupled longitudinal modes are beyond the existing analytic description, but can be quantitatively described by a phenomenological dispersion relation.

Given that hydrodynamic interactions exist in both systems, it is particularly interesting in future studies to evaluate the similarities and, like-wise, the differences in the collective dynamics of the two systems presented here. One obvious difference, particularly important in dense systems as we have considered here, is that the flow perturbation is markedly different in the short range leading to different dynamics as also confirmed by Ref. 40. A complete analytical treatment of this case is an outstanding challenge. In the dilute limit, however, the interactions are dominated by the long-range perturbations and the existence of long-range orientational order in populations of passively flowing droplets coupled via hydrodynamics has recently been shown, both experimentally and analytically, in such systems.[41] An interesting question is whether similar statistical features of the orientational order might show up in collections of active swimmers and how the inherent differences in the energy dissipation might modify the resulting collective dynamics apart from the long-range hydrodynamic coupling.

References

1. T. Vicsek and A. Zaferiris, *Phys. Rep.* **517**, 71–140 (2012).
2. T. Beatus, T. Tlusty and R. Bar-Ziv, *Nat. Phys.* **2**(11), 743–748 (2006).
3. T. Beatus, R. Bar-Ziv and T. Tlusty, *Prog. Theor. Phys. Supp.* **175**, 123–130 (2008).
4. T. Beatus, T. Tlusty and R. Bar-Ziv, *Phys. Rev. Lett.* **103**, 114502 (2009).
5. T. Beatus, R. H. Bar-Ziv and T. Tlusty, *Phys. Rep.* **516**(3), 103–145 (2012).
6. J.-B. Fleury, U. D. Schiller, S. Thutupalli, G. Gompper and R. Seemann, *New J. Phys.* **16**, 063029 (2014).
7. N. Desreumaux, J.-B. Caussin, R. Jeanneret, E. Lauga and D. Bartolo, *Phys. Rev. Lett.* **111**, 118301, (2013).
8. R. Seeman, M. Brinkmann, T. Pfohl and S. Herminghaus, *Rep. Prog. Phys.* **75**, 016601–016642 (2012).
9. S. Thutupalli, R. Seemann and S. Herminghaus, *New J. Phys.* **13**(7), 073021 (2011).
10. T. Brotto, J.-B. Caussin, E. Lauga and D. Bartolo, *Phys. Rev. Lett.* **110**, 038101 (2013).
11. N. Champagne, E. Lauga and D. Bartolo, *Soft Matter* **7**, 11082 (2011).
12. S. Thutupalli and S. Herminghaus, *Eur. Phys. J. E* **36**(8), 1–10 (2013).
13. S. Ramaswamy, *Ann. Rev. Cond. Mat.* **1**, 323 (2010).
14. C. Priest, S. Herminghaus and R. Seemann, *Appl. Phys. Lett.* **88**(2), 024106 (2006).
15. J. G. Santiago, S. T. Wereley, C. D. Meinhart, D. J. Beebe and R. J. Adrian, *Exp. Fluids* **25**(4), 316–319 (1998).
16. W. Thielicke and E. Stamhuis (2005). PIVlab — time-resolved particle image velocimetry (PIV) tool.
17. F. D. Dos Santos and T. Ondarçuhu, *Phys. Rev. Lett.* **75**, 2972–2975 (1995).
18. Y. Hayashima, M. Nagayama, Y. Doi, S. Nakata, M. Kimura and M. Iida, *Phys. Chem. Chem. Phys.* **4**, 1386–1392 (2002).

19. Y. Sumino, H. Kitahata, K. Yoshikawa, M. Nagayama, Sh. M. Nomura, N. Magome and Y. Mori, *Phys. Rev. E* **72**, 041603 (2005).
20. Y.-J. Chen, Y. Nagamine and K. Yoshikawa, *Phys. Rev. E* **80**, 016303 (2009).
21. N. J. Suematsu, Y. Miyahara, Y. Matsuda and S. Nakata, *J. Phys. Chem. C* **114**, 13340–13343 (2010).
22. M. D. Levan and J. Newman, *AIChE J.* **22**(4), 695–701 (1976).
23. M. Levan, *J. Colloid Interf. Sci.* **83**(1), 11–17 (1981).
24. M. T. Downton and H. Stark, *J. Phys.: Condens. Matter* **21**(20), 204101 (2009).
25. A. N. Zaikin and A. M. Zhabotinsky, *Nature* **225**, 535–537 (1970).
26. A. T. Winfree, *Science* **175**, 634–636 (1972).
27. J. P. Hernandez-Ortiz, Ch. G. Stoltz and M. D. Graham, *Phys. Rev. Lett.* **95**, 204501 (2005).
28. A. Sokolov, I. S. Aranson, J. O. Kessler and R. E. Goldstein, *Phys. Rev. Lett.* **98**, 158102 (2007).
29. T. Ishikawa and T. J. Pedley, *Phys. Rev. Lett.* **100**, 088103 (2008).
30. F. Ginelli, F. Peruani, M. Baer and H. Chate, *Phys. Rev. Lett.* **104**, 184502 (2010).
31. V. Narayan, S. Ramaswamy and N. Menon, *Science* **317**, 105–108 (2007).
32. T. Ishikawa, M. P. Simmonds and T. J. Pedley, *J. Fluid Mech.* **568**, 119–160 (2006).
33. I. O. Götze and G. Gompper, *Phys. Rev. E* **82**, 041921 (2010).
34. D. Bensimon, L. P. Kadanoff, S. Liang, B. I. Shraiman and C. Tang, *Rev. Mod. Phys.* **58**, 977–999 (1986).
35. H. Diamant, *J. Phys. Soc. Jpn.* **78**, 041002 (2009).
36. T. Tlusty, *Macromolecules* **39**, 3927–3930 (2006).
37. T. Beatus, R. Bar-Ziv and T. Tlusty, *Phys. Rev. Lett.* **99**, 124502 (2007).
38. N. Desreumaux, N. Florent, E. Lauga and D. Bartolo, *Eur. Phys. J. E* **35**, 68 (2012).
39. I. O. Götze and G. Gompper, *Phys. Rev. E* **84**, 031404 (2011).
40. A. Zöttl and H. Stark, *Phys. Rev. Lett.* **112**, 118101 (2014).
41. I. Shani, T. Beatus, T. Tlusty and R. Bar-Ziv, *Nat. Phys.* (2014).

Chapter 9

MODELING STIMULI-INDUCED RECONFIGURATION AND DIRECTED MOTION OF RESPONSIVE GELS

Debabrata Deb*, Pratyush Dayal[†], Anna C. Balazs*
and Olga Kuksenok*,[‡]

*Department of Chemical Engineering,
University of Pittsburgh, Pittsburgh, PA 15261, USA
[†]Department of Chemical Engineering,
Indian Institute of Technology, Gandhinagar 382424, India
[‡]olk2@pitt.edu

1. Introduction

When responsive gels are exposed to variations in the thermal, optical, or chemical properties of the surrounding medium, these materials can undergo distinct changes in volume and shape, which in turn can be harnessed to create a range of systems with biomimetic functionalities.[1,2] Notably, the ability to dynamically change shape in response to variations in the environment is a distinctly biological attribute that is necessary for the survival of many biological species.[3,4] An analogous ability to undergo dynamic reconfiguration in response to external stimuli could be advantageous for functional materials so that the same material could be molded via different external cues into various shapes and hence have different functionalities. Moreover, the expansions and contractions exhibited by thermo-[5] or chemo-responsive[6,7] gels can be used as synthetic "muscles" in small-scale devices.[5-7] In this review, we describe our recent computational studies that allowed us to model the behavior of responsive gels and gel-based materials in the presence of various stimuli (e.g., light, chemical gradients, and spatially localized exothermic chemical reactions).

We first focus on gels functionalized with spirobenzopyran (SP) chromophores.[8,9] The incorporation of these chromophores into gels formed from N-isopropylacrylamide (PNIPAAm) in aqueous solutions provides a means of harnessing light to control the gels swelling or shrinking.[8,9] We show that these systems can undergo photo-induced shape-reconfiguration, as well as photo-directed motion.[10]

We then turn our attention to a composite system where elastic, catalyst-bearing posts are embedded into thermo-responsive gels.[5] This composite structure enables a cyclic inter-conversion of chemical and mechanical energy that regulates the temperature of the system. Namely, when the catalyst comes in contact with these reagents, an exothermic chemical reaction takes place.[5] The swelling and deswelling of the gel in response to the changes in temperature induces the reversible actuation of the filaments into and out of the reactant-containing layer, switching "on" and "off" chemical reactions.[5] Researchers used various exothermic catalytic reactions to demonstrate different examples of autonomous, self-sustained oscillations in such composite systems.[5]

Finally, in the last section we focus on the behavior of the self-oscillatory gels where the autonomous oscillations of the gels are due to the well known self-oscillatory Belousov–Zhabotinsky (BZ) chemical reaction.[11,12] Namely, we focus on polymer networks encompassing grafted ruthenium catalysts and undergoing the BZ reaction[6,7,13] and examine effects of light on the dynamics of these gels. We also examine interactions between multiple BZ gels and show that this system exhibits a distinct form of chemotaxis, where individual gel pieces respond to chemical gradients and self-organize into a variety of dynamic structures.[14] Notably, these findings can lead to design rules for creating assemblies that effectively communicate to perform a concerted function.

2. Modeling Responsive Gels and Gel-Based Composites

2.1. *Photo-induced reconfiguration and directed motion of spirobenzopyran-functionalized gels*

We first focus on modeling photo-sensitive polymer gels containing spirobenzopyran chromophores, which are anchored onto the polymer network. In the absence of light and in acidic aqueous solutions, the spirobenzopyran chromophores are primarily in the open ring form (the protonated merocyanine form or McH) and are hydrophilic.[8,9,15] Illumination with blue light causes the isomerization of these chromophores into the closed ring conformation (the spiro form or SP), which is hydrophobic.[8,9] In the dark, the SP form is unstable and undergoes spontaneous conversion back to the stable, hydrophilic McH form. Notably, the physical origin of the photo-induced volume change of the SP-functionalized gels is distinctly different from the gel collapse resulting from a direct light-induced heating.[16,17] We recently developed a computational model[10] that captures the dynamics of SP-containing gels. Where possible, the simulation values in our studies[10] were chosen based on the available experimental data[8,9] and our results show good agreement with these experimental findings.[8,9] Specifically, using this model we probed the effect of light on the volume phase transition in SP-containing gels.

Figure 1A shows the equilibrium degree of swelling, $\lambda_{eq}(\tilde{c}_{SP}, T)$, normalized by its value in the dark at $T = 20°C$. We plot $\tilde{\lambda}_{eq} = \lambda_{eq}/\lambda_{eq}(0, 20)$

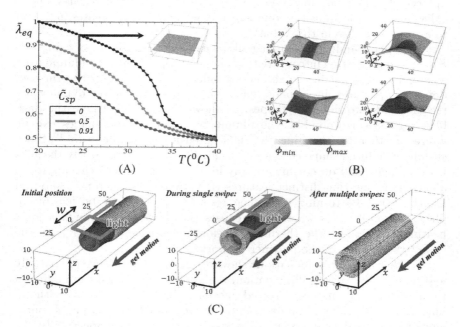

Fig. 1. (A) Volume phase transition in SP-functionalized gels in the dark (black curve) and under illumination (red and blue curves). Inset shows sample in the dark at $T = 25°C$. (B) Sample under non-uniform illumination through photomasks with different apertures; illuminated regions are shown in dark blue. The color shows the volume fraction of polymer; in the color bar, $\phi_{\min} = 0.06$ and $\phi_{\max} = 0.21$. (C) Repeated motion of the illuminated region over the tubular sample in the x-direction (red arrow) results in the sample's motion in the opposite direction (blue arrow). The size of the sample is $50 \times 70 \times 4$ nodes, and the width of the stripe of light is $w = 20$. The rest of the parameters are the same as in Ref. 10.

as a function of temperature for two values of the photo-stationary concentration of chromophores in the SP form, $\tilde{c}_{SP}$ (red and blue lines) and in the absence of light (black line). Note that for the same total concentration of chromophores, higher values of $\tilde{c}_{SP}$ at a given temperature can be achieved by using a higher light intensity so that the blue line in Fig. 1A corresponds to a higher light intensity than the red line. The solid lines in Fig. 1A represent exact analytical solutions while the symbols represent simulation data points;[10] the excellent agreement between the analytical solution and simulation data indicate the accuracy of our simulation approach. Figure 1A shows that under illumination, $\tilde{\lambda}_{eq}$ decreases and the phase transition is significantly smoother than that for the non-illuminated sample, similar to the behavior observed in experiments.[8] Furthermore, we also found that similar to the corresponding experimental findings,[8] the relative light-induced shrinking, $\delta\lambda$, increases with increases in T until it reaches a peak value at the phase transition temperature, while further increases in T lead to a sharp decrease in this value.[10]

Such photo-induced shrinking allows us to dynamically reconfigure this SP-functionalized gel into a variety of shapes.[10] Specifically, by illuminating a sample that is flat in the absence of light (see inset in Fig. 1A) through different apertures in a photomask, the material can be dynamically reconfigured into different shapes, as shown in Fig. 1B, where the dark blue regions correspond to the illuminated areas. These shapes can be altered further by changing the light intensity or temperature.[10] Since such light-induced shape changes are achieved by modifying solely the gel's hydrophilicity and do not involve any local heating effects,[8] this approach allows us to inscribe distinct features that are on the micron to sub-mm length scales. Patterning with such resolution would not be feasible in photo-responsive gels where light causes the deswelling through a local heating[16,17] because the thermal diffusion is significantly faster than the collective diffusion of the polymer network.

Furthermore, our simulations showed that in the presence of spatially and temporally varying illumination, the photo-sensitive gel can be driven to undergo autonomous, directed motion (Fig. 1C). Starting with an initially non-illuminated tubular sample, we repeatedly moved a light source along the positive x-direction, as marked by the red arrow. Specifically, the sample is illuminated through a rectangular aperture in a photomask (the aperture of width $w = 20$ is indicated by the red rectangle in Fig. 1C) and the light is rastered over the sample along the x-direction with a speed v_L. Light causes the region directly below the aperture to shrink; when the light is moved along the sample, the region that is now in the dark swells, and the newly illuminated region shrinks. This swelling/shrinking of contiguous regions results in the movement of the sample in the direction that is opposite to the motion of the light.[10] The gel moves due to the inter-diffusion of the solvent and polymer.[18–21] As the light is moved along the sample, a region of contraction propagates along the sample and effectively "pushes" the solvent in the same direction, and thereby causes the polymer to move to the opposite direction;[10] we will return to the discussion of this mode of motion in the following sections where we focus on the motion of BZ gels. While the gels displacement during a single pass of the light over the sample is small, after multiple passes of the light, the gel undergoes a relatively large net displacement in the negative x-direction (Fig. 1C). This light-induced motion is robust and is observed for a wide range of parameters; it can be controlled by altering the velocity of the light source, light intensity, or the temperature of the sample.[10] Our results point to a robust method for using light to controllably reconfigure the shapes of polymer gels and drive the self-organization of multiple reconfigurable pieces into complex architectures.

2.2. *Self-oscillations in gels-based composites immersed into binary fluid*

We now focus on a distinctly different type of chemo-responsive system: a thermo-responsive gel layer with embedded elastic filaments whose ends are coated with a catalyst for an exothermic chemical reaction. The system is immersed into a binary fluid: the gel with embedded elastic filaments lies below the fluid interface (marked by the red plane in Fig. 2A). The catalyst-coated tips of the filaments can extend into the upper fluid layer, which lies above this plane and contains reagents. When the catalyst comes in contact with these reagents, an exothermic chemical reaction takes place.[5] The swelling and deswelling of the gel in response to the temperature changes drives the reversible actuation of the filaments into and out of the reactant-containing layer.[5]

We emphasize that the gel layer in this system is not chemo-responsive. We introduce a row of elastic filaments at given locations within the gel (see Fig. 2A). Initially (at $T_0 = 22°C$) all the filaments are straight (along the z-direction), with the catalyst-coated tips reaching into the upper reactive layer. We assume that the position of the plane separating the two fluids remains constant.[5] We extended the three-dimensional gLSM model[22] in order to capture the elastodynamics of the gel-based composite with embedded elastic filaments. In addition, in our model we account for the exothermic reaction and the temperature increase when the tips are located above this plane and for the heat dissipation throughout the whole system.[5] Finally, we assume that the temperature changes instantaneously and uniformly across the sample.

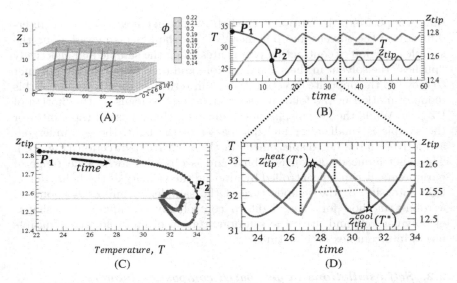

Fig. 2. Self-sustained oscillations in a gel with embedded elastic filaments (shown in (A)). The color bar indicates the volume fraction of polymer, ϕ. Interface between the reactive and non-reactive layers is marked by the red plane. (B) Phase trajectory of the z-coordinate of the tip, $z_{tip}(t)$. (C), (D) Time evolution of z-coordinate of tip position, $z_{tip}(t)$, (blue curve, right axis) and temperature $T(t)$ (red curve, left axis); green line marks the position of the interface (red plane in (A). The image is reproduced with permission from Ref. 5.

When the filaments extend into the reactive layer, they turn on the exothermic chemical reaction.[5] Due to the heat generated by this reaction, the thermo-responsive gel shrinks, causing the filaments to bend and the tips to dip below the reagents layer. Once the tips are removed from the upper layer, the heating stops and the system cools down, causing the gel to expand. The gels expansion drives the filaments to straighten so that the tips move through the plane into the reagent-filled layer, once again giving rise to the exothermic reaction. Figure 2B shows that the phase trajectory of the z-coordinate of the tips, $z_{tip}(T)$, develops into a limit cycle and stays in this cycle indefinitely, corresponding to robust, self-sustained oscillations; here, the point marked P_1 indicates the tips' initial height at 22°C, and P_2 is where the tips first cross the interface.

The time evolution of $z_{tip}(t)$ and $T(t)$ (Figs. 2C–2D) clearly show that the temperature of the system decreases when the tips are below and increases when they are above the plane separating the fluids. Furthermore, at any temperature $T(t)$ within the cycle, the tips are higher when the gel shrinks and lower when it swells, as marked by $z_{tip}^{heat}(T^*) > z_{tip}^{cool}(T^*)$ in Fig. 2D.

This bistability, as well as the negative feedback provided by the localized reaction, results in the oscillations seen in Fig. 2. Bistability plays a key role in various self-oscillating gels.[23-26] The dynamic behavior shown in Figs. 2B-2D is in good qualitative agreement with the experimental findings in Ref. 5. Our model predicted that the oscillation amplitude and period could be controlled by varying the position of the fluid–fluid interface and the heating rate; these trends were confirmed by the corresponding experimental studies.[5] Overall, the system exhibits a remarkable form of chemo-mechano-chemical transduction where chemical and mechanical energy are inter-converted in a continuous, autonomous manner. We emphasize that the chemical reaction in the upper layer is non-oscillatory; in the following two sections, we focus on gels that exhibit oscillatory behavior due to the well known self-oscillatory Belousov–Zhabotinsky (BZ) chemical reaction.[11,12]

2.3. *Stimuli-induced motion in gels undergoing Belousov–Zhabotinsky (BZ) chemical reaction*

The chemo-responsive gels we focus on in this section are undergoing the oscillating chemical reaction known as the Belousov–Zhabotinsky (BZ) reaction.[11,12] These BZ gels encompass a ruthenium catalyst that is covalently bonded to the polymers.[13,27-30] The BZ reaction generates a periodic oxidation and reduction of this anchored catalyst,[13,27-30] which in turn induces a periodic expansion and contraction of the gel due to the hydrating effect of the oxidized catalyst. The ability of BZ gels to transduce chemical energy into mechanical oscillations in the absence of external stimuli[6,7,27,28,30-42] makes them unique materials that are potentially suitable for multiple biomimetic applications. In other words, the periodic chemical oscillations arising from the BZ reaction fuel the mechanical oscillations of these active gels. Hence, in a solution of BZ reagents, the gels swell and deswell autonomously, beating like a heart. Millimeter sized pieces of the BZ gels can oscillate for hours[28,30] and the system can be "resuscitated" by replenishing the solution with reagents that were consumed in the reaction.[43]

In this section, we first focus on the motion of a single BZ gel under non-uniform illumination and then consider multiple chemo-responsive BZ gels immersed in solution and show that the individual gels can "communicate" by sending and receiving a chemical signal, which gives rise to large-scale collective behavior. To model the photo-induced motion of a single gel,[44] we use our three-dimensional gLSM approach,[22] which captures the dynamic evolution of the volume fraction of polymer, ϕ, and the concentrations of the following two critical components: the oxidized catalyst, v, and the

activator, u. To model communication between multiple BZ gels, we use
a gLSM model that is combined with a finite difference approach, which
captures the reaction-diffusion processes for the activator u in the outer
solution[45] and the diffusive exchange of u between the gels and outer
solvent.[45] We account for the effects of light on the BZ gels through the
parameter Φ, which is proportional to the light intensity.[44]

2.3.1. *Negative phototaxis of BZ gel*

In our study of BZ gels under non-uniform illumination, the intensity profile
of the applied light is a step function, so that one-third of the gel is initially
located in the dark and two-thirds is illuminated with a constant intensity.
The gel lies along the x-direction, and the gray shading in Fig. 3A indicates
the region that is in the dark ($\Phi = 0$). Within the illuminated region, we
set $\Phi = \Phi_c$. With $\Phi_c = 3.2 \times 10^{-4}$ the oscillations are suppressed in the
uniformly illuminated sample.[44,46]

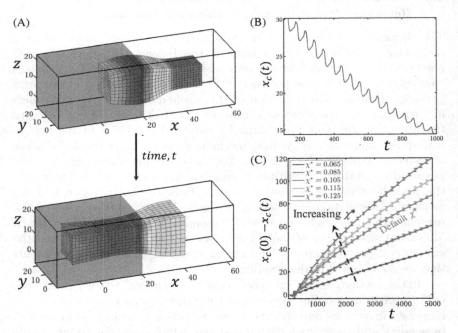

Fig. 3. (A) Motion of BZ gel (of size $30 \times 10 \times 10$ nodes) under non-uniform illumination, snapshots are taken at early ($t = 1200$) and late ($t = 9850$) times. The dark region
is shaded in gray. (B) Evolution of the x-coordinate of the gels center shows locomotion
in the negative x-direction. (C) Deviation of the coordinate of gels center from its initial
position, $(x_c(0) - x_c(t))$, as a function of time.

For the chosen parameters, the sample is in the oscillatory regime in the dark ($\Phi = 0$) and thus, a chemo-mechanical wave is generated at the non-illuminated, left end and travels towards the right. Figure 3A shows the respective morphology and position of the BZ gel at early (top panel) and late (bottom panel) times; the gel is drawn within a larger box to more clearly illustrate the path of its motion. The corresponding trajectory of the x-coordinate of the center of the gel, $x_c(t)$, along with the small scale oscillations due to the periodic swelling and deswelling of the gel, displays a systematic decrease, revealing the net motion of the gel along the negative x-direction (see Fig. 3B). This example illustrates the sample's motion away from the light (negative phototaxis). Namely, spatially non-uniform illumination generates traveling waves from one end of the gel and, therefore, effectively "pushes" the solvent to the right in Figs. 3A–3B. Hence, due to polymer-solvent inter-diffusion, we observe net displacement of the gel that is opposite in direction to the propagation of the traveling waves.[44]

The velocity of the gels directed motion can be tailored by varying χ^*, the parameter that characterizes the responsiveness of the gel.[47] The plots in Fig. 3C show the displacement of the samples at different values of χ^* (but identical illumination) as a function of time. We calculate the change in the x-coordinate of the gel's center, $x_c(t)$, with respect to its initial value, $x_c(0)$, and show that the gel's net velocity increases with increasing χ^*[44] (Fig. 3C). Hence, the negative phototactic motion critically depends on the chemo-mechanical response of the polymer matrix and χ^* must be sufficiently large in order to observe distinct directed motion.

Notably, recent experimental studies have confirmed the above predictions on the photo-controlled autonomous motion of BZ gels. Epstein *et al.*[48] considered a freely sliding BZ gel confined in a capillary tube; when one end of the tube was illuminated with light of a higher intensity than the other end of the tube, the gel displayed negative phototaxis, moving toward the darker region provided that, for the selected intensities, the frequency of oscillations is higher in the darker than in the brighter region.[48]

2.3.2. *Photo-assisted communication between BZ cilia*

To investigate the potential for regulating the interactions among multiple BZ gels via light, we focus on five rectangular gels that are anchored to a substrate and immersed in solvent (see Fig. 4). The cilia are arranged along the x-direction; the value of the activator u is taken to be zero at the sides of the simulation box, whereas periodic and no-flux boundary conditions are imposed in the y and z directions, respectively.[45] Figures 4A–4C show the snapshots of the systems late-time behavior in the absence of light, while Figs. 4D–4F show late-time snapshots of the same array when cilia

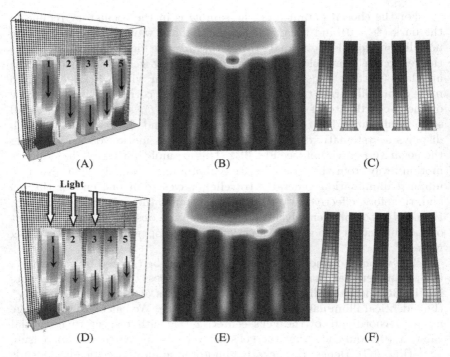

Fig. 4.   Dynamics of the five cilia system. (A)–(C) Late time snapshots in the absence of light ($\Phi = 0$). (D)–(F) Late time snapshots when cilia 1,2 and 3 are illuminated by light ($\Phi = 1.5 \times 10^{-3}$). The size of the simulation box is $69 \times 20 \times 23$ units. The distance between outer cilia surfaces and the fluid boundaries is 5.5 units; the distance between the inner surfaces of the cilia is 3.5 units. The cilium size is $6 \times 6 \times 30$ nodes.

1, 2, and 3 (as marked by arrows in Fig. 4D) are illuminated. Here, $\Phi = 1.5 \times 10^{-3}$, which is greater than the critical value that suppresses the oscillations.[44,49] Hence, under these conditions, an isolated cilium would not exhibit chemo-mechanical oscillations. In Fig. 4D, however, the non-illuminated cilia (numbered 4 and 5) continue to produce the activator u, which diffuses through the fluid to cilia 1, 2, and 3. As the chemical waves travel from right to left (in the negative x-direction), the local activator concentration increases sharply and, thereby, switches "on" each gel. Once the concentration of u becomes sufficiently high within this system, the waves travel synchronously from the top to the bottom of the cilia length (as marked by arrows in Fig. 4D)[45] and the leading wave front shifts towards the non-illuminated side.

Figure 4 reveals that without illumination, the highest concentration of u in the outer fluid occurs above cilium number 3 (Fig. 4B) due to the symmetric arrangement of gels; hence, the outer cilia are bent towards this central unit (Fig. 4C). In the presence of the non-uniform illumination (Fig. 4D), the highest concentration of u is shifted toward cilium number 4 (Fig. 4E) and the cilia are all now titled toward the latter cilium (see Fig. 4F). These results indicate that the cilia move toward the highest concentration of u in the system. This behavior is even more pronounced in systems involving the non-anchored, mobile gels, as we discuss in the following section.[45]

This behavior can be explained as follows. For a small, isolated sample, the intrinsic oscillation frequency of the BZ gel, ω, increases with the concentration of u in the surrounding solution.[45] (Notably, ω in the presence of light remains lower than that in the absence of light for a wide range of u.[45]) In a system of oscillators, the one beating with the highest frequency sets the directionality of wave propagation.[50,51] Hence, regions of each cilium closest to the highest concentration of u have higher intrinsic frequency[45] and set the directionality of the traveling chemical waves in each of the gels. Correspondingly, the gels move in the opposite direction due to the inter-diffusion of polymer and solvent[14,45] as discussed in the section above. In other words, the gels move toward the highest concentration of u in the outer solution; since the bottom surfaces of the cilia are attached to the substrate, this motion results in the "bunching" of the cilia observed in Figs. 4C and 4F. As noted above, ω is higher for a non-illuminated sample; this explains the shift in the position of the leading front towards the cilia in the dark (see Fig. 4D), as well as the asymmetric "bunching" of the cilia towards the non-illuminated region (Fig. 4F).

2.3.3. *BZ gels forming self-rotating "pinwheels"*

In the previous section, we considered BZ gel cilia anchored to a hard substrate and demonstrated how light can be used to control the interaction between these tethered gels. In this section, we again focus on communication between the BZ gels but now we allow the gels to slide freely over the substrate. We have shown that under these conditions BZ gels migrate to the highest concentration of u — a distinct form of autochemotaxis, and this chemotactic behavior can be controlled by exploiting the photosensitivity of the BZ gels.[44] Here, we focus on the scenario where multiple pieces of these gels can effectively "communicate" and collectively harness the chemical energy from the BZ reaction to self-organize into self-rotating "pinwheels" or gears.[52]

The system of interest is shown in Fig. 5, where we consider four equally spaced BZ gel cubes that are placed at the bottom of the simulation box and are able to slide freely on this surface. The simulation box is $49 \times 49 \times 13$

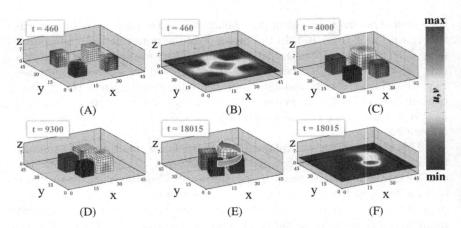

Fig. 5.   Mechanism for forming a pinwheel. (A), (C), (D), and (E) show the positions of four self-oscillating BZ gels at times $t = 460$ (A), 4000 (C), 9300 (D) and 18015 (E). The color represents the concentration of oxidized catalyst, v, in the gels. The red arrow in (E) marks the pinwheels direction of rotation. (B) and (F): The distribution of activator concentration, u, in the $z = 6$ plane of the simulation box at times $t = 460$ (E) and 28035 (F) (corresponding to gel positions in (A) and (E)). Blue (red) represents the minimum (maximum) value of u and v and the color range used is as indicated in the color bar on the right.

units (~ 1.23 mm $\times$ 1.23 mm $\times$ 0.33 mm) in size and each BZ cube is $l = 8.9$ units with the initial center-to-center separation between two neighboring gels equal to 20 units. We set $u = 0$ along all the side walls of the simulation box and apply no-flux boundary conditions on the top and bottom walls. As the reaction precedes, the activator, u, produced within the polymer network, diffuses into the surrounding fluid, and due to the symmetric arrangement of the gels', becomes highly concentrated between the four gels (see Fig. 5B); this localized high concentration of u results in the gels motion to the center (see Figs. 5C and 5D).[52] The physical reason for the gels motion towards the higher concentration of u was discussed in the previous section for the case of BZ cilia. While the same effect resulted in the "bunching" of the gels anchored to surface, the freely sliding gels are able to move significantly closer together,[14] setting the stage for the new phenomena described here: the autonomous formation of self-rotating "pinwheels".

In what we refer to as a "pinwheel" (Fig. 5E), the four-cube assembly rotates around a central point (in a counterclockwise direction in this example). Furthermore, each cube rotates around its center of mass (also in a counterclockwise direction). The concurrent temporal evolution of u resembles a moving "comet", with a large "head" and a trailing "tail"

(Fig. 5F), which circles around the center of the four-cube cluster in the opposite direction of the moving gels (i.e., in a clockwise direction).

The formation of such a self-rotating "pinwheel" occurs in two stages.[52] In the first stage, the four gels move closer to each other toward the center of the box and at a particular instant in time, the chemo-mechanical oscillations among the gels become phase locked. At this time, we observe a constant phase difference between the oscillations in neighboring gels and consequently, the chemical wave travels successively from one gel to the next (in a clockwise direction in Fig. 5). This clockwise motion of the chemical wave results in the counterclockwise motion of the gels. During the second stage, gels continue to move toward each other until the distance between the outer surfaces of neighboring gels reaches a critical distance of $r_c = 1.5$ units, when the excluded volume interactions begin to affect the gels motion[52] and create a torque, resulting in the rotation of both the entire cluster and the individual gels. We refer to the latter instant of time as the time of pinwheel formation. In this particular example (Fig. 5), the first stage begins at $t = 4000$, while the second stage begins at $t = 9300$.

The above example represents the outcome of one simulation; for another independent simulation using a different random seed, the system did not undergo rotation, but simply exhibited auto-chemotaxis, with the gels coming together and remaining in a configuration similar to Fig. 5D. To gain further insight into the factors leading to the rotary motion, we compare the time evolution of quantities characterizing both the rotating and non-rotating assemblies in Fig. 6, where the points marked (a)–(e) correspond to images in Fig. 5. Figure 6 shows that both the average concentration of activator, u_{avg}, and the oscillation frequency, ω, are significantly higher for the rotating than for the non-rotating assembly. This observation is consistent with our previous study that showed an increase in the oscillation frequency, ω, of an isolated gel with the increase in the concentration of u in the outer solution.[45]

We emphasize that both rotating and non-rotating scenarios are outcomes of the same simulation set-up but with different initial random fluctuations, indicating that the system in Fig. 5 is bistable. Moreover, the probability of realizing the non-rotating assembly is significantly higher than that of forming a pinwheel: out of 24 independent simulations, we observed pinwheel formation in only three cases. However, Fig. 6 provides clues to the factors that promote the formation of pinwheels by indicating that a critical difference between the rotating and non-rotating assemblies is the higher average concentration of u in the rotary system. Indeed, we have shown that by increasing the concentration of u in the simulation box (by imposing a high concentration of u at the center of the top wall), the

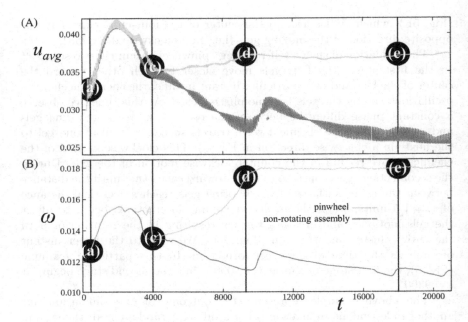

Fig. 6. Time evolution of quantities related to the pinwheel (green curves) and non-rotating assembly (red curves). (A) Evolution of the average activator concentration, u_{avg}. (B) Evolution of the frequency of oscillations, ω, in one of the four gels. The points marked as (a)–(e) on the green curves correspond to the respective systems shown in the Fig. 5(a)–(e).

probability of forming the rotating assembly could be significantly increased and that a non-rotating assembly can also be turned into a pinwheel.[52]

In the above scenario, the directionality of the pinwheels rotation depended only on initial random fluctuations. We have also shown that this directionality can be controlled by applying gradients in u at the four walls of the simulation box.[52] Figure 7 shows the evolution of a system that formed a pinwheel rotating in the counterclockwise direction as we changed the boundary conditions in multiple stages. (Notably, without this change in the boundary condition, the same system would evolve to a non-rotating assembly). During Stage 1, we set a linear gradient in u from $u = 0$ to 0.6 along all the four edges of the simulation box in a counterclockwise manner (see right panel in Fig. 7E, where the distribution of u in the plane at $z = 6$ is shown at $t = 110$). Here, Stage 1 lasts from $t = 0$ to 14000, during which all the gels move relatively far apart from each other towards the high u at the boundaries so that the distance between the cubes' centers is roughly 20 units by $t = 14000$ (i.e., similar to the initial distance between

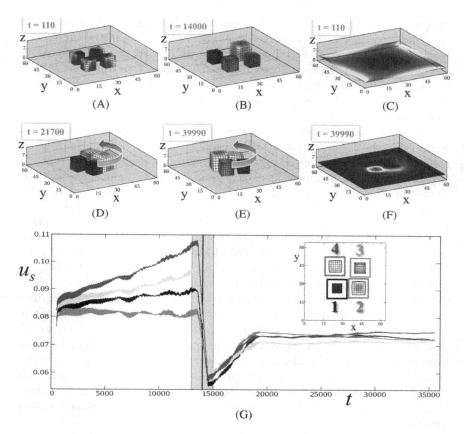

Fig. 7. Controlling the direction of rotation of a pinwheel. Stage 1: ($t < 14000$) gradients in u from $u = 0$ to 0.6 is set at the four edges of the simulation box; see (A) for gel positions and (E) for u at $z = 6$ at $t = 110$). Stage 2: from $t = 14001$ when $u = 0$ is set at four edges of simulation box and ends at $t = 21700$ (see (B) for gel positions at $t = 14000$). Stage 3 corresponds to gels forming pinwheel at $t > 21700$ (see (D) for gel positions and (F) for u at $z = 6$). (G) The running average of u_s, within a shell of size $d_s = 16$ around each BZ gel. The vertical black line marks $t = 14000$ when the boundary condition is reset to $u = 0$. Inset shows positions of the gels at $t = 14000$ and the rectangles around the gels mark the dynamic shells where we calculate u_s; the colors of the curves correspond to the numbering of the gels indicated in the inset.

the gels in the example in Fig. 5). Interestingly, at the end of Stage 1, the gel pieces have not only been driven away from each other, but also they are now spaced slightly asymmetrically with respect to each other,[52] and hence, the average value of u around each gel is different from its neighbor's. In effect, the counterclockwise linear gradient in u has served to break the symmetry of the system in Fig. 7A; now the gel surrounded by the highest concentration of u can act as the pacemaker. With this initial influx of u having served this symmetry breaking role, we now set $u = 0$ at the edges at time $t = 14001$. During Stage 2 (Fig. 7C) (from $t = 14001$ to 21700), the system effectively replicates processes in Figs. 5A–5C. Finally, during Stage 3 ($t > 21700$), the gels form a pinwheel (Fig. 7D) rotating in the counterclockwise direction.

To quantify how application of the gradient of u at the boundary in Fig. 7 effectively controls the average distributions of u around the gels, we measured the total activator, u_s, contained within a dynamic shell (cube of size $d_s = 16$ units encompassing each gel and solvent around it, as shown by the rectangles around gels in Fig. 7G). Running averages of u_s over time for the four gels are shown in Fig. 7G. For the counterclockwise rotation, the plot clearly demonstrates that the value of u_s around gel 4 is the highest at the time when the boundary condition is switched to $u = 0$ by the end of Stage 1 ($t = 14000$), with the second highest value being around gel 3 (see inset for gel numbering). Such distributions of u_s at the end of Stage 1 sets the directionality of the rotating traveling wave to be in the clockwise direction (from gel 4 to gel 3), which results in a pinwheel rotating in a counterclockwise direction. The introduction of a comparable gradient in the clockwise direction during Stage 1 results in a pinwheel rotating in the clockwise direction.[52] In other words, changing the patterning on the wall from counterclockwise to clockwise effectively "flips" the average values of u around these gels and thereby their direction of rotation. Furthermore, this process is robust and the clockwise (counterclockwise) pinwheel formation is now observed in all these simulations (18 independent runs for each direction).

Finally, we consider an assembly of 16 cubes as shown in Fig. 8. The box is now $124 \times 124 \times 13$ units in size and the gel pieces are placed within four clusters in the box (see Fig. 8A). The initial center-to-center distance between the clusters is 70 units; within each cluster, the initial center-to-center distance between the gels is 25 units. In this scenario, each of the clusters forms a pinwheel. We observed that the system self-organized into four separate pinwheels in four out of five independent simulations; in one case, the gels organized into three pinwheels and one non-rotating assembly. We attribute this significantly higher probability of pinwheel formation in the larger system to the higher asymmetry in the positioning of the gels at later times; this in turn leads to differences in u_s around the gels within each cluster, thereby, promoting pinwheel formation.[52] Our results indicate that the motion induced by the self-generated chemical gradients

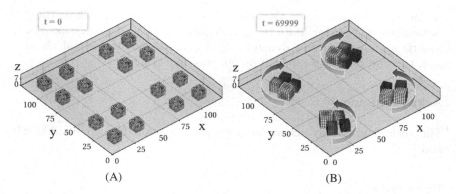

Fig. 8. Dynamics in colony of 16 self-oscillating BZ gels at times $t = 0$ (A), 69990 (B). Arrows mark the direction of rotation of the pinwheels.

and the excluded volume interactions give rise to a novel dynamic state: namely, the four cubes of BZ gels self-organize into a pinwheel.

3. Conclusion

Through computational modeling, we uncovered a range of biomimetic behavior in responsive gels and gel-based composites. Specifically, we demonstrated that photo-sensitive SP-functionalized gels can undergo both dynamic reconfiguration and directed motion. The ability to remotely and non-invasively alter the shape of SP-containing gels and manipulate their transport could potentially be useful for bringing multiple samples together and driving them to "dock" to form soft, self-assembled structures with remotely reconfigurable architectures.

We also demonstrated a remarkable form of chemo-mechano-chemical transduction in a thermo-responsive gel that contained catalyst-coated filaments. These findings can guide the development of "smart" switches for regulating chemical reactions and maintaining the temperature in microfluidic devices. Finally, we focused on self-oscillating BZ gels and showed that these systems can move in response to such stimuli as light or chemical gradients. While single BZ gel samples undergo negative phototaxis (move away from light), multiple gels are capable of self-organizing into both non-rotating and rotating assemblies. Hence, these materials could form simple self-propelled machines, such as gears, that perform autonomous work. Overall, the studies indicate that these computational models can facilitate the development of the next generation of soft robotic systems that can dynamically respond to environmental changes and, thereby, exhibit a range of biomimetic functionality.

Acknowledgments

The authors gratefully acknowledge financial support from ARO, AFOSR and DOE.

References

1. M. A. C. Stuart, W. T. S. Huck, J. Genzer, M. Mller, C. Ober, M. Stamm, G. B. Sukhorukov, I. Szleifer, V. V. Tsukruk, M. Urban, F. Winnik, S. Zauscher, I. Luzinov and S. Minko, *Nat. Mater.* **9**(2), 101–113 (2010).
2. R. Geryak and V. V. Tsukruk (2013). Reconfigurable and actuating structures from soft materials, *Soft Matter.*
3. M. D. Norman, J. Finn and T. Tregenza, *Proc. Biol. Sci.* **268**(1478), 1755–1758 (2001).
4. J. J. Allen, L. M. Mthger, A. Barbosa and R. T. Hanlon, *J. Comp. Physiol. A* **195**(6), 547–555 (2009).
5. X. He, M. Aizenberg, O. Kuksenok, L. D. Zarzar, A. Shastri, A. C. Balazs and J. Aizenberg, *Nature* **487**(7406), 214–218 (2012).
6. R. Yoshida, *Bull. Chem. Soc. Jpn.* **81**(6), 676–688 (2008).
7. S. Maeda, Y. Hara, T. Sakai, R. Yoshida and S. Hashimoto, *Adv. Mater.* **19**(21) 3480–3484 (2007).
8. A. Szilgyi, K. Sumaru, S. Sugiura, T. Takagi, T. Shinbo, M. Zrnyi and T. Kanamori, *Chem. Mater.* **19**(11), 2730–2732 (2007).
9. T. Satoh, K. Sumaru, T. Takagi and T. Kanamori, *Soft Matter* **7**(18), 8030–8034 (2011).
10. O. Kuksenok and A. C. Balazs, *Adv. Func. Mater.* **23**(36), 4601–4610 (2013).
11. B. P. Belousov, *Collection of Short Papers on Radiation Medicine* (Medgiz, Moskow, 1959), pp. 145–152.
12. A. N. Zaikin and A. M. Zhabotinsky, *Nature* **225**(5232), 535–537 (1970).
13. R. Yoshida, T. Takahashi, T. Yamaguchi and H. Ichijo, *J. Am. Chem. Soc.* **118**(21), 5134–5135 (1996).
14. P. Dayal, O. Kuksenok and A. C. Balazs, *PNAS* (2012), p. 201213432.
15. T. Satoh, K. Sumaru, T. Takagi, K. Takai and T. Kanamori, *Phys. Chem. Chem. Phys.* **13**(16), 7322–7329 (2011).
16. A. Suzuki and T. Tanaka, *Nature* **346**(6282), 345–347 (1990).
17. A. Suzuki, Phase transition in gels of sub-millimeter size induced by interaction with stimuli, in K. Duek (ed.), *Responsive Gels: Volume Transitions II*, no. 110 in Advances in Polymer Science (Springer Berlin Heidelberg, 1993), ISBN 978-3-540-56970-1, 978-3-540-47836-2, pp. 199–240.

18. V. V. Yashin and A. C. Balazs, *Science* **314**(5800), 798–801 (2006).
19. V. V. Yashin and A. C. Balazs, *J. Chem. Phys.* **126**(12), 124707 (2007).
20. J. Boissonade, *Chaos* **15**(2), 023703 (2005).
21. T. Roose and A. C. Fowler, *Bull. Math. Biol.* **70**(6), 1772–1789 (2008).
22. O. Kuksenok, V. V. Yashin and A. C. Balazs, *Phys. Rev. E* **78**(4), 041406 (2008).
23. J.-C. Leroux and R. A. Siegel, *Chaos* **9**(2), 267–275 (1999).
24. R. A. Siegel, *Autonomous Rhythmic Drug Delivery Systems Based on Chemical and Biochemomechanical Oscillators*, NATO Science for Peace and Security Series A: Chemistry and Biology (Springer, Dordrecht, The Netherlands, 2009), p. 200.
25. J. Boissonade, *Phys. Rev. Lett.* **90**(18), 188302 (2003).
26. J. Horvth, I. Szalai, J. Boissonade and P. D. Kepper, *Soft Matter* **7**(18), 8462–8472 (2011).
27. R. Yoshida, S. Onodera, T. Yamaguchi and E. Kokufuta, *J. Phys. Chem. A* **103**(43), 8573–8578 (1999).
28. S. Sasaki, S. Koga, R. Yoshida and T. Yamaguchi, *Langmuir* **19**(14), 5595–5600 (2003).
29. R. Yoshida, *Colloid Polym. Sci.* **289**(5), 475–487 (2011).
30. K. Miyakawa, F. Sakamoto, R. Yoshida, E. Kokufuta and T. Yamaguchi, *Phys. Rev. E* **62**(1), 793–798 (2000).
31. R. Yoshida, E. Kokufuta and T. Yamaguchi, *Chaos* **9**(2), 260–266 (1999).
32. T. Sakai and R. Yoshida, *Langmuir* **20**(4), 1036–1038 (2004).
33. R. Yoshida, M. Tanaka, S. Onodera, T. Yamaguchi and E. Kokufuta, *J. Phys. Chem. A* **104**(32), 7549–7555 (2000).
34. R. Yoshida, *Cur. Org. Chem.* **9**(16), 1617–1641 (2005).
35. Y. Murase, S. Maeda, S. Hashimoto and R. Yoshida, *Langmuir* **25**(1), 483–489 (2009).
36. S. Shinohara, T. Seki, T. Sakai, R. Yoshida and Y. Takeoka, *Angew. Chem. Int. Ed.* **47**(47), 9039–9043 (2008).
37. J. Shen, S. Pullela, M. Marquez and Z. Cheng, *J. Phys. Chem. A* **111**(48), 12081–12085 (2007).
38. S. Tateyama, Y. Shibuta and R. Yoshida, *J. Phys. Chem. B* **112**(6), 1777–1782 (2008).
39. S. Maeda, Y. Hara, R. Yoshida and S. Hashimoto, *Angewandte Chem.* **120**(35), 6792–6795 (2008).
40. S. Maeda, Y. Hara, R. Yoshida and S. Hashimoto, *Chaos* **29**(5), 401405 (2008).
41. D. Suzuki and R. Yoshida, *Macromolecules* **41**(15), 5830–5838 (2008).
42. D. Suzuki and R. Yoshida, *J. Phys. Chem. B* **112**(40), 12618–12624 (2008).
43. I. C. Chen, O. Kuksenok, V. V. Yashin, R. M. Moslin, A. C. Balazs and K. J. V. Vliet, *Soft Matter* **7**(7), 3141–3146 (2011).
44. P. Dayal, O. Kuksenok and A. C. Balazs, *Langmuir* **25**(8), 4298–4301 (2009).
45. P. Dayal, O. Kuksenok, A. Bhattacharya and A. C. Balazs, *J. Mater. Chem.* **22**(1), 241–250 (2011).
46. P. Dayal, O. Kuksenok and A. C. Balazs, *Soft Matter* **6**(4), 768–773 (2010).

47. V. V. Yashin and A. C. Balazs, *Macromolecules* **39**(6), 2024–2026 (2006).
48. X. Lu, L. Ren, Q. Gao, Y. Zhao, S. Wang, J. Yang and I. R. Epstein, *Chem. Commun.* **49**(70), 7690–7692 (2013).
49. O. Kuksenok, P. Dayal, A. Bhattacharya, V. V. Yashin, D. Deb, I. C. Chen, K. J. V. Vliet and A. C. Balazs, *Chem. Soc. Rev.* **42**(17), 7257–7277 (2013).
50. A. S. Mikhailov and A. Engel, *Phys. Lett. A* **117**(5), 257–260 (1986).
51. O.-U. Kheowan, E. Mihaliuk, B. Blasius, I. Sendia-Nadal and K. Showalter, *Phys. Rev. Lett.* **98**(7), 074101 (2007).
52. D. Deb, O. Kuksenok, P. Dayal and A. C. Balazs, *Mater. Horiz.* **1**(1), 125–132 (2014).

Chapter 10

DISSIPATIVE BZ PATTERNS IN SYSTEMS
OF COUPLED NANO- AND MICRODROPLETS

Vladimir K. Vanag

Institute for Chemistry and Biology,
Immanuel Kant Baltic Federal University,
A.Nevskogo str. 14A, Kaliningrad, 236041, Russia
vkvanag@gmail.com

Irving R. Epstein

Department of Chemistry, Brandeis University,
Waltham, MA 02454-9110
epstein@brandeis.edu

1. Introduction

Modern nonlinear chemistry arguably began with the work of Belousov and Zhabotinsky,[1,2] who first studied their famous oscillatory chemical reaction in a stirred reactor. Such an oscillatory reaction, when it takes place in a continuously stirred tank reactor (CSTR) is a so-called point (or 0-dimensional = 0D) oscillator. There are now many such oscillatory reactions known,[3,4] for example, the CDIMA reaction,[5,6] pH-oscillators,[7,8] bromate-sulfite-ferrocyanide,[9] the Briggs–Rauscher reaction,[10] etc. The mathematics of oscillatory reactions is based on sets of ordinary differential equations and the theory of nonlinear dynamics, which include such key notions as bifurcation analysis, the Hopf theorem, limit cycles and Poincaré sections.

If we pour a solution of the BZ reactants into a Petri dish to form a thin layer and remove any stirring (to eliminate advection), we obtain a homogeneous spatially extended dissipative system that can be described by a set of reaction-diffusion equations (partial differential equations). The theory of dissipative patterns in reaction-diffusion systems was developed by Turing,[11] Prigogine,[12] Zaikin and Zhabotinsky,[13] Winfree[14] and others. Basically, and very roughly, spatially extended patterns can be understood in terms of two diffusive instabilities, the Turing and wave instabilities,[11] as well as excitability, which is associated with the phenomenon of propagating trigger waves.[13,15] Turing and wave instabilities can be characterized by linear stability analysis of the linearized reaction-diffusion equations. In the case of Turing instability, the largest eigenvalue is positive in some range of wave numbers, k, provided that the imaginary part of this eigenvalue is equal to zero over this same range of k. Turing instability leads to temporally stationary patterns periodic in space. In the case of wave instability, the imaginary part of the largest eigenvalue (whose real part is positive in some range of k) is not equal to zero. This gives patterns periodic both in space and time, e.g., standing waves. The first experimental evidence for Turing patterns was obtained using the CDIMA reaction,[5] while chemical standing waves were obtained first in the BZ-AOT system,[16] which we describe later.

Another way to add complexity to a nonlinear chemical system is to take several 0D oscillators and to couple them somehow. A system of coupled oscillators can be described by a (possibly quite large) number of ordinary differential equations.[17] The prototype for such systems in nature is a neural network (ensemble of neurons). Mathematically, such a system of coupled oscillators does not have a spatial dimension.

In reality, however, homogeneous reaction-diffusion systems are not completely homogeneous, and systems of coupled oscillators extend in space, so that their behavior should depend on the real geometrical configuration of oscillators and on the spatial dimension of the system. In other words, real systems are heterogeneous reaction-diffusion systems. To study experimentally such systems, we have developed two similar but different configurations. The first is the BZ-AOT system,[18] which is closer to homogeneous reaction-diffusion systems; the second is an array of coupled microdroplets with the BZ reaction taking place inside each aqueous droplet,[19] which is closer to a system of coupled point oscillators.

First, we briefly describe the BZ reaction. Then we explain the BZ-AOT and the BZ microdroplet systems. Finally, we compare mechanisms of pattern formation in these two systems.

2. The Belousov–Zhabotinsky Reaction

The BZ reaction is the metal ion (or metal ion complex)-catalyzed bromination and oxidation of an organic substrate, usually malonic acid, in a strongly acidic medium.[1,2] The reaction in a closed system can persist in the oscillatory regime (with a period of 1–5 min) for hundreds of cycles. This property makes the BZ reaction unique in nonlinear chemical dynamics. If the catalyst is ferroin [tris(1,10-phenanthroline)iron(II)], or bathoferroin [tris(4,7-diphenyl-1,10-phenanthroline)iron(II)], or $Ru(bipy)_3^{2+}$ [tris(2,2'-bipyridyl)ruthenium(II)], then the BZ system can be easily monitored spectrophotometrically, since the system oscillates between the reduced and oxidized states of the catalyst, which have very different absorption maxima. The catalyst $Ru(bipy)_3^{2+}$ (even at relatively low concentrations as a co-catalyst) makes the BZ reaction photo-sensitive, allowing for perturbation and control by light. The BZ reaction displays an astonishing variety of periodic and chaotic temporal oscillations, as well as trigger waves (typically spirals or target patterns). All these features make the BZ reaction the prototype system for studies of nonlinear chemical dynamics.[1,20,21] The mechanism of the BZ reaction is well described by the FKN model[22] and its simplified version, the three-variable Oregonator model proposed by Field and Noyes.[23]

3. The BZ-AOT System

The general idea that compartmentalization might play a key role in the origin of life and in evolution[24] makes it appealing to include or encapsulate chemical oscillatory reactions in small interacting droplets. If the BZ reaction takes place in a large number of distinct but interacting volumes (water droplets) under conditions where some signaling species can diffuse between these volumes, then we can obtain a system with synchronized nano- or micro-oscillators. Such a system can be created by using well-known water-in-oil microemulsions that consist of a mixture of water, oil and surfactant. A thermodynamically stable reverse microemulsion of water nanodroplets in oil is easily created if the surfactant is the ionic surfactant AOT [aerosol OT, sodium bis(2-ethylhexyl)sulfosuccinate].[25] Since the BZ reactants and catalyst are polar, they are essentially confined to the aqueous nanodroplets, though non-polar intermediates such as Br_2 and BrO_2 (which are generated during the reaction) may enter and diffuse in the continuous oil phase, providing a means of interdroplet communication.

There is also a second means of communication between nanodroplets: a process of collision followed by fusion-fission, which is relatively fast for neighboring droplets. The characteristic time t_{ex} of this process is roughly $(k_{ex} \times c_d)^{-1}$, where the rate of mass exchange[26] $k_{ex} \approx 10^7 \ M^{-1} s^{-1}$ and c_d is the concentration of nanodroplets. At $c_d \approx 10^{-3} - 10^{-4} \ M$, $t_{ex} \approx 1 \ ms$ or less.

The polar head groups of the AOT molecules project in toward the water core, while the non-polar tails are directed out into the oil. The average radius of the water core is determined by the ratio of water to surfactant molecules, $\omega = [H_2O]/[AOT]$, and is given in nm by $R_w \approx 0.17\omega$.[27] Since we typically work with mixtures in which ω lies in the range 9–25, we are dealing with droplets that are several nanometers in diameter. The spacing between droplets is controlled by the volume fraction, φ_d, of the dispersed phase (volume fraction of water, φ_w, plus volume fraction of surfactant, φ_{AOT}, $\varphi_w + \varphi_{AOT} + \varphi_{oil} = 1$, where φ_{oil} is the volume fraction of oil).

Many physical properties of the AOT microemulsion depend on φ_d. This dependence is due in large part to percolation. As φ_d increases, clusters of droplets begin to grow,[28] and at a critical value, φ_p (around 0.5), percolation takes place, resulting in the formation of long dynamic channels of water. In Fig. 1, we show schematically a single water nanodroplet and computer simulations of clustering in an AOT microemulsion below the percolation transition.[29]

Below the percolation threshold, the diffusion coefficient, D_d, of a water droplet and its radius, R_d, are related by the Stokes–Einstein formula

$$D_d = kT/(6\pi\eta R_d)$$

where η is the viscosity of the organic solvent, k is the Boltzmann constant, and T is the temperature. The droplet's diffusion coefficient and the diffusion coefficients of water soluble molecules (which diffuse together

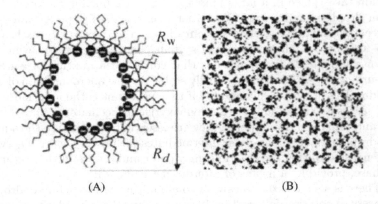

(A)   (B)

Fig. 1. (A) A water droplet of the AOT-microemulsion. R_w is the radius of the droplet's water core, R_d is the outer radius of the droplet, equal to the hydrodynamic radius measured experimentally by dynamic light scattering. (B) Computer simulation[29] of an AOT microemulsion. Each black dot represents a single water nanodroplet; droplet fraction $\varphi_d = 0.3$. Size is 1 mμ by 1 mμ.

with water droplets) are one or two orders of magnitude (depending on the droplet radius and ϕ_d) smaller than the diffusion coefficient of small molecules in the oil phase, which is approximately $10^{-5}\,\mathrm{cm^2\,s^{-1}}$.[30] Above percolation, the diffusion coefficient of water-soluble molecules increases by one to two orders of magnitude[30] due to the free diffusion of these molecules through the dynamical water channels.

Communication between droplets takes place via two distinct diffusive processes. Highly polar species (e.g., ions such as Br$^-$), which are largely confined to the water droplets, are exchanged when droplets collide and undergo fission and fusion. Less polar (HBrO$_2$) or non-polar (Br$_2$) species diffuse through the oil as single molecules and display diffusion coefficients typical of single molecules.

By varying the structure of the AOT microemulsion, i.e., the droplet fraction φ_d and the radius R_w of the droplets, as well as the chemistry of the BZ system, which can be characterized by the ratio of initial concentrations [H$_2$SO$_4$][NaBrO$_3$]/[MA] (a rough measure of the relative strengths of activation and inhibition in the reaction mixture) and the catalyst, a large gallery of patterns can be obtained (see Fig. 2). Most of the phenomena shown in Fig. 2 are unique to the BZ-AOT system; they have not been observed in experiments on the BZ reaction in simple aqueous solution.

At droplet fraction $\phi_d < \phi_p$ (below percolation), temporally stationary patterns induced by Turing instability are observed: Turing patterns (spots and stripes or labyrinthine patterns) and localized stationary spots ("chemical memory"). In a narrow range of parameters where sub-critical Turing and sub-critical Hopf instabilities interact, oscillons, i.e., localized oscillatory spots, were found.[31]

At larger droplet fraction or with other (larger and more hydrophobic) catalysts (like bathoferroin), which induce formation of nanodroplet clusters, or at higher temperature (which also leads to clustering and percolation), new types of wave patterns emerge. Some of these waves, like standing waves or packet waves (in which groups of traveling waves move through space coherently) arise from a wave instability. Others, like antispirals (in which traveling spiral waves propagate inward toward the spiral core rather than outward from it), accelerating waves (which accelerate before collision with an incoming wave), bubble waves (which consist of localized spots such that each spot does not move forward but emerges in synchrony with neighboring spots, grows and disappears), segmented spirals (in which waves split into small segments transverse to the direction of propagation), and jumping waves (in which circular waves propagate outward in a discontinuous, saltatory fashion) may have different interpretations including wave instability as well.

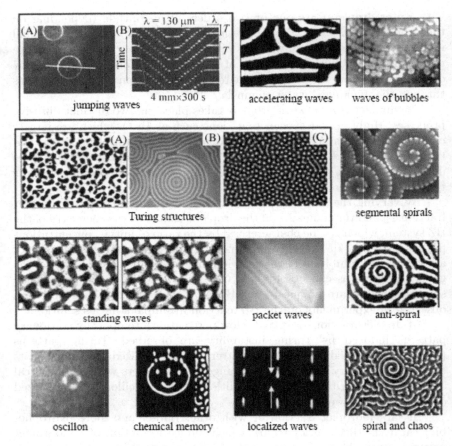

Fig. 2. Patterns found in the BZ-AOT system. Jumping waves (first row) are presented as snapshots (A) and as a space-time plot [(B), obtained across the white line shown in (A)]. Two other snapshots present accelerating waves and waves of "bubbles". Second row: Three types of Turing patterns: (A) "black dots" (honeycomb), (B) "frozen waves" and (C) "white dots" (distorted hexagons). The last snapshot shows segmented spirals. Third row: Standing waves are shown as two antiphase snapshots. Packet waves and antispirals (as well as standing waves) demonstrate patterns induced by wave instability. 4th row: Oscillon, "chemical memory" or localized stationary spots and localized waves demonstrate different types of localized patterns. "Chemical memory" is obtained by brief illumination through a mask of a photosensitive BZ system. Localized waves are obtained in the same manner using a striped mask. The last snapshot shows co-existence of spiral waves and chaotic waves. Typical size of all snapshots is $5 \pm 2\,\text{mm}$.

The BZ-AOT system is a remarkable prototype system for pattern formation also due to its ability to support 3D patterns. We have utilized tomographic techniques to study three-dimensional Turing patterns and 3D standing waves in the BZ-AOT system.[32,33] Examples of such 3D patterns are shown in Figs. 3 and 4.

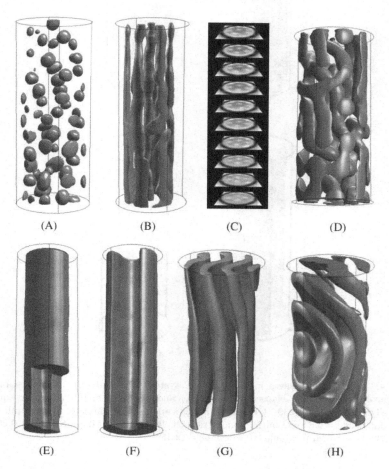

Fig. 3.   Tomographically reconstructed concentration fields for 3D Turing patterns. Spots (A); hexagonally packed cylinders (B); horizontal cross-sections (C) taken through the data array in (B); labyrinthine pattern (D); tube (E); half-pipe (F); lamellar pattern (G); and concentric hemispherical lamellae (H). [Malonic acid]$_0$ = 0.30 M; [H$_2$SO$_4$]$_0$ = 0.18 M, [NaBrO$_3$]$_0$ = 0.23 M, [ferroin] = 5 mM, inner diameter of capillary is 0.6 mm for (A)–(D), (G), (H) and 0.3 mm for (E), (F). A short segment at the bottom of the front half of pattern (E) has been removed to reveal the inner structure.

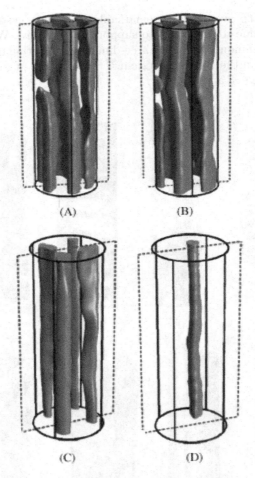

Fig. 4.   3D standing waves in the BZ-AOT system. (A) and (B): lamellar standing waves, $\phi_d = 0.46$. Time elapsed between (A) and (B) is 26 s. (C) and (D): square-packed cylinders, $\phi_d = 0.39$. (C) and (D) are 28 s apart. Inner diameter of the capillary = 0.3 mm; height = 0.8 mm. Initial concentrations: $[MA]_0 = 0.3$ M, $[H_2SO_4]_0 = 0.25$ M, $[NaBrO_3]_0 = 0.3$ M, $[ferroin]_0 = 0.4$ mM; $\omega = 12.35$.

The theory for Turing instability in reaction-diffusion systems requires that the diffusion coefficient for the inhibitor be larger or even much larger than the diffusion coefficient for the activator. In the BZ system, $HBrO_2$ serves as the activator, while Br^- and/or Br_2 are inhibitors. Compartmentalization at the nano level in a water-in-oil microemulsion results in quite different diffusion coefficients for the activator, which is mostly inside the nanodroplets, and the inhibitor species, Br_2, which diffuses through the oil phase. Note also that a nanodroplet is too small to be considered as a single oscillator. Indeed, a catalyst molecule is found on average in only about one of ten nanodroplets. The characteristic wavelength of the patterns shown in Fig. 2 is about 0.2 mm, 100000 times larger than the diameter of a single nanodroplet, of which there are about 10^{17} in a typical experiment. The characteristic time of mass exchange between droplets, a few milliseconds, is about 10000 times shorter than the period of oscillations. Thus (1) the level of organization observed in our patterns clearly results from the collective interaction of many droplets, and (2) the BZ-AOT system can be considered as a homogeneous system and can be described by partial differential reaction-diffusion equations.

We use several reaction-diffusion models for the BZ-AOT system.[16,18,31,34] These models are based on the modified Oregonator model augmented with species that partition into the oil phase and hence diffuse more quickly than the polar BZ species. All of these models give good qualitative agreement with our experiments. Note also that wave instability requires at least a three-variable model, for example, a model with a water-soluble activator and inhibitor (with small diffusion coefficients for both) and an additional (oil soluble) species that can diffuse much faster than the first two. Wave instability can be found in a rather large range of parameters if this additional species is coupled to the activator. However, coupling to the inhibitor also can give rise to wave instability.

4. BZ Microdroplets

Now consider what happens when the diameter of the water droplets is close to the characteristic size of the observed dissipative patterns. In this case, we can no longer view the system as homogenous and must introduce discreteness in the description of our coupled micro-oscillators. By employing microfluidic techniques, we have been able to obtain one- and two-dimensional droplet arrays with diameters in the range 50–300 μm. These droplets offer several advantages. They are large enough that individual droplets can be seen under a microscope. Unlike the nanodroplets, they do not undergo random Brownian motion, but can be maintained at a fixed position. This enables us, when the system is made photo-sensitive by employing the $Ru(bipy)_3^{2+}$ catalyst, to selectively perturb individual droplets with light. Droplets with larger diameters communicate too weakly to be of interest. In addition, wave patterns emerge in droplets with diameter larger than 300 μm, and consequently such droplets cannot be considered as point (or 0D) oscillators.

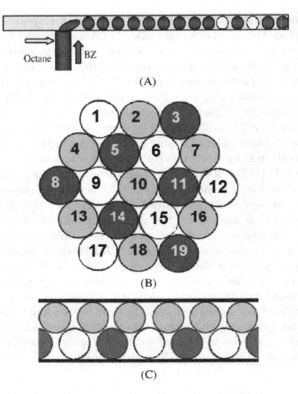

Fig. 5. (A) Schematic rendering of method used to produce BZ droplets with a microfluidic device. Capillary diameter is typically about 100 μm. (B) Schematic of two-dimensional "π-S" pattern. Gray circles represent stationary (non-oscillatory) droplets; black and white circles show oscillatory droplets 180° out of phase with each other. (C) Schematic of "1.5 D" pattern with stationary and oscillatory droplets labeled as in (B).

Figure 5A shows schematically how the droplets are generated by flowing streams of oil and aqueous BZ solution into a narrow capillary. The coupling is dominated by the inhibitory species, Br_2, especially if a very hydrophobic oil (like octane) is used. In one dimension, i.e., in a capillary tube whose diameter is equal to that of the droplets, the most stable pattern is antiphase oscillations, in which neighboring droplets oscillate 180° out of phase.[19] The strength of the coupling depends upon the diameter of and distance between the drops, the malonic acid concentration, and the period of the BZ oscillations. In addition, changing the oil can change the ratio of partition coefficients P between the aqueous and oil phases both for the hydrophobic inhibitor, Br_2, and the hydrophilic activator, $HBrO_2$.

This ratio, as we will show later, can be important for the pattern formation mechanism. At large enough inhibitory coupling strength, stationary Turing patterns, where droplets in the oxidized and reduced steady states alternate in a capillary, have also been obtained.[19]

In two dimensions, the observed patterns can be more elaborate.[35] Figure 5C shows a "1.5 D" pattern, in which the capillary is wide enough to hold two intercalated rows of droplets. In the pattern illustrated, one row of droplets is in a stationary state, while the other row contains droplets oscillating antiphase. In a wider vessel, the droplets self-assemble into a two-dimensional hexagonal lattice, and we observe many stable patterns, including the one illustrated schematically in Fig. 5B, where in each triangle of droplets one is stationary and the other two oscillate antiphase. A fully oscillatory pattern with droplets oscillating 120° out of phase with their neighbors is also stable. Many less symmetric patterns are also obtained.

Turing patterns and antiphase oscillations are found in a system of non-moving BZ microdroplets under the condition that the diffusion coefficients of all species are almost identical both in the oil and aqueous phases. What is the reason for pattern formation in such a system? From theory and experiments with coupled oscillators, we know that inhibitory coupling of two oscillators can lead to oscillator death, a dynamic behavior which is similar to Turing patterns.[36,37] Our own computer simulations of two BZ droplets coupled via the oil phase reveal that stationary Turing patterns occur for pure inhibitory coupling and antiphase oscillations (due to wave instability) can be obtained for pure excitatory coupling.[38]

If we describe the dynamics of coupled BZ droplets in terms of ordinary differential equations, then the coupling strength, C_U, between two BZ droplets (shown schematically in Fig. 6) and the corresponding equations (in the absence of any reaction) can be written in the following form[38,39]:

$$du_1/dt = C_U(s_{12}/P_U - u_1)$$

$$ds_{12}/dt = C_U r_V(u_1 + u_2 - 2s_{12}/P_U)$$

$$du_2/dt = C_U(s_{12}/P_U - u_2)$$

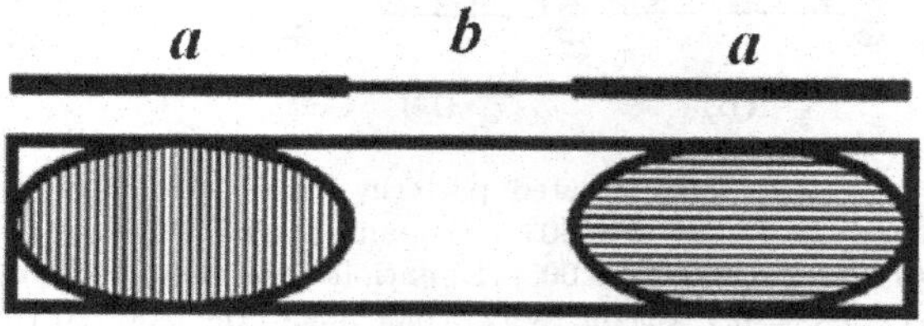

Fig. 6. Two BZ droplets separated by an oil gap (length b). a is the length of a BZ droplet.

where $u_1 (u_2)$ is the concentration of a species U in the left (right) droplet, s_{12} is the concentration of U in the oil phase (between the two droplets),

$$C_U = k_f r_V P_U / (1 + r_V P_U),$$

$k_f = 2D/a^2$, $r_V = a/b$, D is the diffusion coefficient ($\approx 10^{-5} \, \text{cm}^2 \, \text{s}^{-1}$), $P_U = u_1/s_{12} = u_2/s_{12}$ at equilibrium.

If U is the inhibitor, Br_2, then Turing instability for the BZ droplets can occur at a large enough coupling strength. If U is the activator, $HBrO_2$, then wave instability is found with equal diffusion coefficients for activator and inhibitor. This means that the partition coefficients P or, more precisely, the ratio $P_{\text{inhibitor}}/P_{\text{activator}}$ (in a heterogeneous system) play approximately the same role as the ratio $D_{\text{inhibitor}}/D_{\text{activator}}$ for homogeneous systems.

If we take a 1D array of BZ droplets (described by the ordinary differential equations of the Oregonator-like model and locally coupled

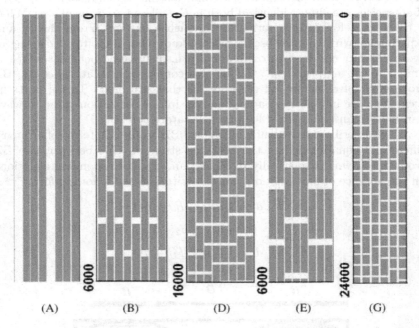

Fig. 7. Space-time plots for simulated patterns of 6–8 circularly coupled droplets. (A) Turing pattern at $P_X = 0$–0.0035; (B) anti-phase oscillation, $P_X = 0$–0.003; (D) moving clusters, $P_X = 0.004 - 0.005$; (E) antiphase oscillatory clusters, $P_X = 0.005$; (G) moving defect, $P_X = 0.1$. Bright (dark) color marks the oxidized (reduced) state of the catalyst. Numbers at the bottom are total time (in seconds) for each run. Coupling parameters: $k_f = 0.45 \, \text{s}^{-1}$, $r_V = 10$, $P_I = 2.5$.

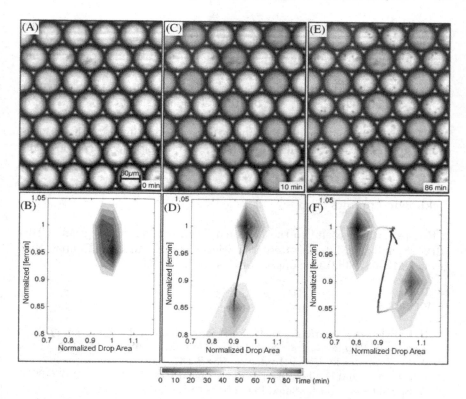

Fig. 8.   Images (upper row) and histograms (lower row) of drops demonstrating morphogenesis.[40] Histograms show fraction of original concentration of reduced BZ catalyst versus fraction of original drop area. Grey scale in histograms is proportional to the number of drops with a given area and intensity. Color coded lines track the center of each peak as a function of time. (A) ,(B) Initially, drops are nearly homogeneous in both intensity and size. (C), (D) At intermediate times, drops undergo a Turing bifurcation, becoming heterogeneous in oxidation state, but remaining homogeneous in size, as seen by the differentiation into lighter and darker drops. (E), (F) Later, drops become heterogeneous in both oxidation state and size. The oxidized (brighter) drops shrink and the reduced drops swell.

via oil-soluble activator and inhibitor signaling species), fix the inhibitory coupling strength (using a constant partition coefficient for the inhibitor, P_I) and only vary the strength of excitatory coupling by changing the partition coefficient of the activator, P_X, then many different patterns can be obtained (see Fig. 7).[39] We conclude that the partition coefficients control the coupling strength, and the balance between inhibitory and excitatory coupling gives a variety of different patterns.

A recent set of experiments[40] on 2D arrays of BZ droplets provides the first experimental evidence for Turing's prediction[11] that reaction and diffusion can result in the differentiation of initially identical cells into populations with *physically different* characteristics. Figure 8 shows a hexagonal array of BZ droplets that first undergo a chemical transformation to a Turing state consisting of a mixture of equally sized droplets, some in a reduced and some in an oxidized steady state. After a bit more than an hour, a second transition, due to osmotic forces, occurs in which the reduced drops grow and the oxidized drops shrink, generating two morphologically different types of "cells."

Acknowledgments

I. R. Epstein is grateful to the National Science Foundation (grant CHE-1012428) and the Materials Research Science and Engineering Center (NSF grant DMR-0820492) for support of this work.

References

1. B. P. Belousov, A periodic reaction and its mechanism. In *Collection of Short Papers on Radiation Medicine* (Medgiz, Moscow, 1959), pp. 145–152.
2. A. M. Zhabotinsky, *Proc. Acad. Sci. USSR* **157**, 392–395 (1964).
3. I. R. Epstein and K. Showalter, *J. Phys. Chem.* **100**(31), 13132–13147 (1996).
4. A. S. Mikhailov and K. Showalter, *Chaos* **18**(2), 0261011 (2008).
5. V. Castets, E. Dulos, J. Boissonade and P. De Kepper, *Phys. Rev. Lett.* **64**(24), 2953–2956 (1990).
6. P. De Kepper, I. R. Epstein, K. Kustin and M. Orbán, *J. Phys. Chem.* **86**(2), 170–171 (1982).
7. V. Gaspar and Showalter, *J. Am. Chem. Soc.* **109**(16), 4869–4876 (1987).
8. G. Rábai, M. Orban and I. R. Epstein, *J. Phys. Chem.* **96**(13), 5414–5419 (1992).
9. E. C. Edblom, M. Orbán and I. R. Epstein, *J. Am. Chem. Soc.* **108**(11), 2826–2830 (1986).
10. S. D. Furrow, *J. Chem. Educ.* **89**(11), 1421–1424 (2012).
11. A. M. Turing, *Philos. Trans. R. Soc. London. Ser. B* **237**, 37–72 (1952).
12. I. Prigogine and R. Lefever, *J. Chem. Phys.* **48**(4), 1695–1700 (1968).
13. A. N. Zaikin and A. M. Zhabotinsky, *Nature* **225**, 535–537 (1970).
14. A. T. Winfree, *Science* **175**, 634–636 (1972).
15. J. Zagora, M. Voslar, L. Schreiberova and I. Schreiber, *Phys. Chem. Chem. Phys.* **4**(8), 1284–1291 (2002).
16. V. K. Vanag and I. R. Epstein, *Phys. Rev. Lett.* **87**(22), 228301 (2001).
17. A. Pikovsky, M. Rosenblum and J. Kurths, *Synchronization. A Universal Concept in Nonlinear Sciences* (University Press: Cambridge, 2003).

18. V. K. Vanag, *Phys. Usp.* **47**(9), 923–941 (2004).
19. M. Toiya, V. K. Vanag and I. R. Epstein, *Angew. Chem. Int. Ed.* **47**(40), 7753–7755 (2008).
20. I. R. Epstein and J. A. Pojman, *An Introduction to Nonlinear Chemical Dynamics* (Oxford University Press, New York, 1998).
21. R. J. Field and M. Burger, *Oscillations and Traveling Waves in Chemical Systems* (Wiley, New York, 1985).
22. R. J. Field, E. Körös and R. M. Noyes, *J. Am. Chem. Soc.* **94**(25), 8649–8664 (1972).
23. R. J. Field and R. M. Noyes, *J. Chem. Phys.* **60**(5), 1877–1884 (1974).
24. O. J. Miller, K. Bernath, J. J. Agresti, G. Amitai, B. T. Kelly, E. Mastro-battista, V. Taly, S. Magdassi, D. S. Tawfik, A. D. Griffiths, *Nat. Methods* **3**, 561–570 (2006).
25. T. K. De and A. Maitra, *Adv. Colloid Interface Sci.* **59**, 95–193 (1995).
26. J. Lang, A. Jada and A. Malliaris, *J. Phys. Chem.* **92**(7), 1946–1953 (1988).
27. M. S. Baptista and C. D. Tran, *J. Phys. Chem. B* **101**(21), 4209–4217 (1997).
28. M. Kotlarchyk, S. H. Chen and J. S. Huang, *J. Phys. Chem.* **86**(17), 3273–3276 (1982).
29. V. K. Vanag, *J. Mendeleev Rus. Chem. Soc.* **LIII**(6), 16–24 (2009).
30. L. J. Schwartz, C. L. DeCiantis, S. Chapman, B. K. Kelley and J. P. Hornak, *Langmuir* **15**(17), 5461–5466 (1999).
31. V. K. Vanag and I. R. Epstein, *Phys. Rev. Lett.* **92**(12), 128301 (2004).
32. T. Bansagi, V. K. Vanag and I. R. Epstein, *Science* **331**(6022), 1309–1312 (2011).
33. T. Bansagi, V. K. Vanag and I. R. Epstein, *Phys. Rev. E* **86**(4), 045202 (2012).
34. V. K. Vanag and I. R. Epstein, *J. Chem. Phys.* **131**(10), 104512 (2009).
35. M. Toiya, H. O. Gonzalez-Ochoa, V. K. Vanag, S. Fraden and I. R. Epstein, *J. Phys. Chem. Lett.* **1**(8), 1241–1246 (2010).
36. M. F. Crowley and I. R. Epstein, *J. Phys. Chem.* **93**(6), 2496–2502 (1989).
37. A. Koseska, E. Volkov and J. Kurths, *Phys. Rev. Lett.* **111**(2), 024103 (2013).
38. V. K. Vanag and I. R. Epstein, *J. Chem. Phys.* **119**(14), 7297–7307 (2003).
39. V. K. Vanag and I. R. Epstein, *Phys. Rev. E* **84**, 066209 (2011).
40. N. Tompkins, N. Li, C. Girabawe, M. Heymann, G. B. Ermentrout, I. R. Epstein and S. Fraden, *Proc. Nat. Acad. Sci. USA* **111**(12), 4397–4402 (2014).

Chapter 11

CONTROL OF CHEMICAL WAVE PROPAGATION

Jakob Löber, Rhoslyn Coles, Julien Siebert, Harald Engel, and Eckehard Schöll

Institut für Theoretische Physik,
Technische Universität Berlin,
D-10623 Berlin, Germany

1. Introduction

Besides the well-known Turing patterns, reaction-diffusion (RD) systems possess a rich variety of spatio-temporal structures.[1-3] Spatially one-dimensional examples include traveling fronts, solitary pulses, and periodic pulse trains that are the building blocks of more complicated patterns in two- and three-dimensional active media as, e.g., spiral turbulence and scroll waves, respectively. Another important class of RD patterns forms stationary, breathing, moving or self-replicating localized spots. Labyrinthine patterns as well as phase turbulence, defect-mediated spiral turbulence and scroll wave turbulence are examples for more complex patterns. In the Belousov–Zhabotinsky (BZ) reaction in microemulsions the BZ inhibitor (bromide) is produced in nanodroplets and diffuses through the oil phase at a rate up to two orders of magnitude greater than that of the BZ activator (bromous acid). In this heterogeneous RD system, a variety of patterns including three-dimensional Turing patterns have been observed by computer tomography (see Ref. 4 and references therein).

Several control strategies have been developed for the purposeful manipulation of RD patterns. Below, we will differentiate between closed-loop or feedback control with and without nonlocal spatial coupling[5-7] or time delay,[8,9] and open-loop control that includes external spatio-temporal forcing, optimal control,[10] control by imposed geometric constraints or heterogeneities, and others.[11-14] While feedback control relies on continuously running monitoring of the system's state, open-loop control is based on a detailed knowledge of the system's dynamics and its parameters.

Feedback-mediated control has been applied quite successfully to the control of propagating one-dimensional (1D) waves as well as to spiral waves in 2D that are among the most prominent examples of spatio-temporal patterns in oscillatory and excitable RD systems. Crucial for the control of spiral waves dynamics is the resonant drift of the spiral core in response to a periodic change of the medium's excitability exactly at the spiral's rotation frequency.[15] Under external resonant periodic forcing, the drift direction depends on the orientation of the spiral wave as there is no synchronization between the externally applied control signal and the intrinsic spiral wave dynamics. In contrast, with feedback-mediated parametric forcing, the direction of resonant drift is independent of the spiral's current orientation. Therefore, we can assign a unique vector to each point of the plane specifying the absolute value and the direction of feedback-induced resonant drift imposed on a spiral core located at the given position.[16] Stable fixed points and stable limit cycles are the possible attractors of the drift velocity field while separatrices of saddle points and unstable limit cycles form the boundaries of the basins of attraction. Thus, from the drift velocity field we gain complete information about the asymptotic motion of the spiral core under feedback control. Given that, first, feedback-induced drift remains small, and, second, the shape of the spiral wave can be approximated by an Archimedian spiral, from the RD equations for the concentration fields, ordinary differential equations can be derived for the temporal variation of the core center coordinates. Different control loops have been realized in experiments with the photo-sensitive BZ reaction using feedback signals obtained from wave activity measured at one or several detector points, along detector lines, or in a spatially extended control domain including global feedback control. Possible control parameters range from the gain and the time delay in the feedback loop over the detector position to the size and the geometrical shape of the control domain. The theoretical predictions agree well with the experimental data, for details see for example Refs. 17, 18. Furthermore, feedback-mediated control loops can be employed in order to stabilize unstable patterns, such as unstable traveling wave segments and spots. This was shown in experiments with the photo-sensitive BZ reaction.[19] Two feedback loops were used to guide unstable wave segments along pre-given trajectories.[20]

Under feedback control, a meandering spiral wave can be forced to rotate rigidly in a parameter range where rigid rotation is unstable in the absence of feedback.[21]

An open loop control was successfully deployed in dragging traveling chemical pulses of adsorbed CO during heterogeneous catalysis on platinum single crystal surfaces.[22] In these experiments, pulse velocity was controlled by a laser beam creating a movable localized temperature heterogeneity on an addressable catalyst surface.[23-25] Dragging a chemical front or phase interface to a new position by anchoring it to a movable parameter heterogeneity, was studied theoretically in Refs. 26, 27, 28.

Many complex RD patterns can be understood as composed of interfaces, fronts, solitary excitation pulses, etc. Before tackling control of complex patterns, it makes sense to develop a detailed understanding of the control of these simpler "building blocks". We choose the Schlögl model as a particularly simple, to some extent analytically tractable example of front dynamics in bistable RD media.[29-32] In Sec. 2, experimentally feasible options are discussed for manipulating front dynamics. The effect of nonlocal feedback on front propagation is analyzed in Sec. 3, while Sec. 4 presents an analytically derived open loop control tailored to a precise control of the front position over time.

2. The Schlögl Model

2.1. *The Schlögl model as an autocatalytic reaction mechanism*

In 1972, Schlögl discussed the autocatalytic trimolecular RD scheme[29,30]

$$A_1 + 2X \underset{k_1^-}{\overset{k_1^+}{\rightleftharpoons}} 3X, \qquad\qquad X \underset{k_2^-}{\overset{k_2^+}{\rightleftharpoons}} A_2 \qquad (1)$$

as a prototype of a non-equilibrium first order phase transition. Concentrations c_1 and c_2 of the chemicals A_1 and A_2, respectively, are assumed to be kept fixed by appropriate feeding of the continuously stirred open reactor where the reaction takes place. Supply of A_1 and removal of A_2 maintains the RD system far from equilibrium. The reaction kinetics $R(u)$ of the chemical X with concentration u is cubic

$$R(u) = -k_1^- u^3 + k_1^+ c_1 u^2 - k_2^+ u + k_2^- c_2. \qquad (2)$$

Given that in some parameter range $R(u) = 0$ possesses three real non-negative roots, $0 \le u_1 < u_2 < u_3$, kinetics (2) can be cast into the form

$$R(u) = -k(u - u_1)(u - u_2)(u - u_3) \qquad (3)$$

where the constant parameters k and u_i ($i = 1, 2, 3$) can be expressed by the parameters in $R(u)$, Eq. (2). Taking into account diffusion of X, the time evolution of the concentration field $u(x, t)$ is given by the RD equation

$$\partial_t u = D\partial_x^2 u - k(u - u_1)(u - u_2)(u - u_3), \tag{4}$$

where D denotes the diffusion coefficient. Initially, Eq. (4) has been discussed in 1937 by Zeldovich and Frank-Kamenetsky in connection with flame propagation.[33] Under spatially non-uniform conditions, concentrations $c_{1/2}$ are still assumed to be constant in space and time. For an open unstirred reactor fed solely by mass flow through the reactor boundaries, this assumption holds if components $A_{1/2}$ diffuse much faster than component X. A more sophisticated realization for e.g., microfluidic devices would be an application of spatially distributed nozzles within the reactor which continuously exchange the solution such that the concentrations of $A_{1/2}$ are kept constant everywhere.

A linear stability analysis reveals that $u_{1/3}$ and u_2 represent stable and unstable homogeneous steady states (HSS) of the RD system Eq. (4), respectively. Therefore, in a certain parameter range the Schlögl model describes a bistable chemical reaction. In the following, we focus on the control of the narrow interface between two co-existing domains of the stable and metastable phases $u_{1/3}$. Within this interface, concentration u changes rapidly from one stable HSS to the other. The globally stable state tends to invade the entire available space developing an expanding or retracting front that displaces the metastable domain. For the Schlögl model, the front profile $u(x, t) = U_c(z)$ in the co-moving frame $z = x - ct$, and the front velocity, c, are known analytically

$$U_c(z) = \frac{1}{2}(u_1 + u_3) + \frac{1}{2}(u_1 - u_3)\tanh\left(\frac{1}{2}\sqrt{\frac{k}{2D}}(u_3 - u_1)z\right), \tag{5}$$

$$c = \sqrt{\frac{Dk}{2}}(u_1 + u_3 - 2u_2). \tag{6}$$

The front solution $U_c(z)$ connects the upper stable stationary state u_3 for $z \to -\infty$ with the lower stable stationary state u_1 for $z \to \infty$. The front travels to the right as long as $u_1 + u_3 > 2u_2$. A stationary front, i.e., a persistent spatial phase separation, exists only for the parameter combination with

$$u_1 + u_3 = 2u_2 \Rightarrow c = 0. \tag{7}$$

Likewise, there exist front solutions which travel to the left with a negative velocity and satisfy interchanged boundary conditions. Figure 1 (right) shows a space-time plot of uncontrolled front propagation in response to the symmetric initial u-profile displayed in Fig. 1 (left). Two fronts propagate

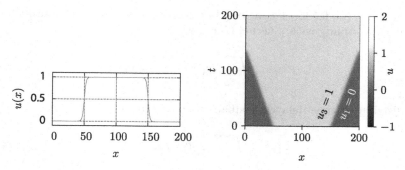

Fig. 1. Space-time plot of two traveling fronts (right) evolving from the symmetric initial concentration profile (left). Numerical solution of the uncontrolled Schlögl equations with periodic boundary conditions posed at the ends of a finite spatial domain. Spatial discretization step: $\Delta x = 0.2$; maximum simulation time $T = 200$; time discretization step: $\Delta t = 0.01$. Parameters $u_3 = 1$, $u_2 = 0.25$, $u_1 = 0$, $k = D = 1$.

at the same speed in opposite directions where the globally stable domain $u = u_3$ (yellow) invades the metastable domain $u = u_1$ shown in red in Fig. 1.

In the following we write the controlled Schlögl model in the general form

$$\partial_t u = D\partial_x^2 u + R(u) + F(u). \tag{8}$$

Here, the control term F is a functional of $u(x,t)$, which may contain time-delayed or nonlocal terms of the concentration variable u, described for instance by integrals over spatial or temporal kernels as discussed in the next section. The control term may also depend upon x, t, denoting an external spatio-temporal control signal. In case of open-loop control, the control signal is prescribed independently of the state of the system. For closed-loop or feedback control, the system state must be monitored, and F turns out to be a functional of u realizing a state-dependent control.

Experimentally, the state u of the controlled system can be measured only with limited accuracy. Often measurements are restricted to a certain region of the spatial domain supporting wave propagation. Furthermore, in multi-component RD systems rarely all components are simultaneously accessible to measurements. A system is called observable in control theory if it is possible to fully reconstruct its state from output measurements.

Subsequently we discuss different options for the control of the RD system Eq. (8) in consideration of physically based constraints and derive the corresponding coupling functions.

First of all, feedback control (closed-loop control) may be described by nonlocal coupling with a kernel $G(x')$, e.g.,

$$F(u) = \int_{-\infty}^{+\infty} G(x')u(x - x')dx' \tag{9}$$

or distributed time-delayed feedback with a kernel $G(t')$

$$F(u) = \int_{0}^{+\infty} G(t')u(t - t')dt'. \tag{10}$$

Second, an additive external forcing $f(x,t)$ that describes time-dependent local concentration sources or sinks leads to $F = f(x,t)$ (open-loop control). In this case f specifies the current local rate at which component X is added to or removed from a reactor. An experimental realization is a nearly continuous array of nozzles diluting or concentrating the solution and thereby changing the concentration of component X at the prescribed rate f. The rate can be positive or negative and is not subject to any explicit restrictions. However, X cannot be removed if its concentration u reaches its minimum possible value $u = 0$. This state constraint applies to any RD system whose components represent chemical concentrations.

Further, control can be realized via parameters in the reaction kinetics. We set $F = \mathcal{G}(u)f(x,t)$ where $\mathcal{G}$ is a possibly state-dependent function that depends on the mechanism by means of which the control signal is coupled to the RD system. If the reaction function $R(u; p)$ depends on some experimentally controllable parameter p, we replace this parameter by $p + \epsilon f(x,t)$. Expanding with respect to the small control amplitude ϵ it follows in leading order

$$\mathcal{G}(u) = \partial_p R(u; p). \tag{11}$$

Assume, for example, that concentration c_2 in the Schlögl model can be controlled spatio-temporally. Then, replacing the constant c_2 in Eq. (2) by the space-time dependent quantity $c_2 + \epsilon f(x,t)$, along the same line of reasoning as before, we end up with a state-independent coupling

$$\mathcal{G} = k_2^-. \tag{12}$$

However, now the explicit control constraint $c_2 + \epsilon f(x,t) > 0$ has to be valid. A similar consideration starting with concentration c_1 leads to a state constraint $c_1 + \epsilon f(x,t) > 0$ and a multiplicative control with state-dependent coupling function

$$\mathcal{G} = k_1^+ u^2. \tag{13}$$

Finally, apart from the concentrations $c_{1/2}$, the reaction rate $R(u)$ Eq. (2) depends on the kinetic coefficients $k_{1/2}^\pm$ that in turn are functions of the temperature T according to the Arrhenius law $k_{1/2}^\pm \sim e^{-E_{1/2}^\pm/(k_B T)}$. Here, k_B and $E_{1/2}^\pm$ denote the Boltzmann constant and the activation energy of the corresponding reaction, respectively. The temperature dependence was exploited for the control of pattern formation in the catalytic CO oxidation on a single crystal Pt(110) surface.[23] In the experiments, a computer-controlled movable laser beam induced a localized temperature heterogeneity from which reaction fronts and pulses nucleated.[25] Gently dragged by the laser beam, pulses were moved over the surface with a velocity up to twice as large as the velocity of the uncontrolled pulse.[22] The temperature field response is very fast compared to the time scale of the catalytic reaction, and the heat released by the reactions is negligible. Modeling the controlled reaction by three coupled RD equations with a laser-induced localized Gaussian temperature heterogeneity of small amplitude ($|T(x,t)| \leq 1K$) and small standard deviation ($2\,\mu m$) qualitatively confirmed the experimental results. To incorporate a temperature-mediated control of the Schlögl model Eq. (8), we substitute T by $T + \epsilon f(x,t)$ and obtain for small ϵ the coupling function

$$\mathcal{G}(u) = \frac{1}{k_B T^2} \left(E_1^- k_1^- u^3 + E_2^+ k_2^+ u - c_1 E_1^+ k_1^+ u^2 - c_2 E_2^+ k_2^- \right). \tag{14}$$

Assuming for simplicity that all activation energies are equal, $E_1^- = E_2^- = E_1^+ = E_2^+ = E$, the coupling function turns out to be proportional to the reaction rate R,

$$\mathcal{G}(u) = -\frac{E}{k_B T^2} R(u) \tag{15}$$

and the controlled Schlögl model reads

$$\partial_t u = D \partial_x^2 u + R(u) - \frac{\epsilon E}{k_B T^2} R(u) f(x,t) \tag{16}$$

subjected to the constraint $T + \epsilon f(x,t) > 0$.

For the Schlögl model with a coupling function $\mathcal{G}(u) \sim R(u)$, the effect of a stationary periodic modulation $f(x)$ on the propagation velocity of traveling front solutions was analytically investigated in Refs. 34 and 35.

In accordance with numerical simulations, a spatially periodic modulation with zero spatial average generally lowers the average propagation velocity. Propagation failure, or pinning, of fronts occurs if the spatial period of the modulation is of the same order as the front width. Outside this interval, i.e., for smaller and larger periods, front propagation is still possible. Thus, in a proper range of spatial periods a temperature modulation results in a persistent spatial phase separation.

Finally, we briefly mention control by non-uniform boundary conditions. This method, not encompassed by the chosen form of a controlled RD system according to Eq. (8), is nevertheless important for applications. Boundary control assumes that the mass exchange flow transfer $J(t)$ of chemical species X can be prescribed at the boundaries such that the boundary conditions in the one-dimensional finite domain $[x_0, x_1]$ are given by

$$\partial_x u\left(x_0, t\right) = J_0(t), \qquad\qquad \partial_x u\left(x_1, t\right) = J_1(t) \qquad (17)$$

with continuously adjustable boundary flows $J_{0/1}(t)$. We refer to Ref. 36 for an example of how a desired stationary concentration profile can be enforced onto a RD system applying boundary control. More difficult approaches based on control signals coupled nonlinearly into the system are possible and experimentally relevant but not considered in this contribution.

2.2. Experimental realizations of the Schlögl model

Examples of isothermal chemical systems exhibiting multiple stationary states are few. The Schlögl model does not describe a realistic reaction scheme because it involves a trimolecular reaction step, and the latter is based on the unlikely reactive collision between three molecules within a small volume. However, due to Korzhukin's theorem, for any homogeneous chemical reaction with polynomial reaction rate one can write down an equivalent set of unimolecular reactions between intermediate species.[37-39] One notable example is the iodate oxidation of arsenous acid.[40] The resulting reaction rate in an RD equation (4) for the concentration of iodide ions $u = \left[I^-\right]$ reads[41]

$$R(u) = (k_1 + k_2 u)(c_1 - u)c_2 u. \qquad (18)$$

Here, $c_1 = \left[IO_3^-\right]$ denotes the concentration of iodate while $c_2 = \left[H^+\right]^2$ where $\left[H^+\right]$ is the concentration of hydrogen ions. In the uncontrolled system concentrations c_1 and c_2 are assumed to be constant. Obviously, both of them, as well as the rate constants $k_{1/2}$ and the temperature could be considered for spatio-temporal control as described above in the case of the Schlögl model.

Another elaborately studied experimental example for chemical bistability is the CO oxidation on Pt(111) single crystal surfaces. This reaction has been modeled by two coupled RD equations for the surface concentrations of adsorbed CO and oxygen. In this case, bistability relies on the Langmuir–Hinshelwood mechanism. Possible control parameters are partial pressures of CO and oxygen in the gas phase.[42] Front propagation has been also observed in experiments with the CO oxidation on Pt(110).[43]

Because of the close similarity of generation and recombination processes of charge carriers in semiconductors with chemical reactions, the Schlögl model can also be applied to self-organization in semiconductors induced by nonlinear generation and recombination processes.[32,44] The analogy of pattern formation in chemical and electrochemical systems,[45] and in particular the front dynamics in bistable semiconductor models,[46] and its control by time-delayed feedback[47,48] has been extensively studied.

Another non-chemical model of bistable dynamics, which is quite interesting from the viewpoint of spatio-temporal control, is a liquid crystal light valve (LCLV) inserted in an optical feedback loop.[49–51] A nematic liquid crystal film is placed between a glass plate and a photoconductive material. Applying an external voltage V_0 across the cell with the help of transparent electrodes orients the polar molecules in parallel to the electric field. A spatio-temporal illumination distribution on the photoconductive wall modulates the electric field locally and allows for spatio-temporal control of the orientation of the liquid crystal molecules. Near to the so-called point of "nascent bistability" the normalized average director $u(x,t)$ obeys the RD equation

$$\partial_t u = D\partial_x^2 u - u^3 + \tilde{\epsilon}u + \eta + (b + du)f(x,t). \qquad (19)$$

Here, f denotes the spatio-temporal forcing signal that is proportional to the applied light intensity, while constants η, $\tilde{\epsilon}$, b and d are related to the properties of the LCLV and depend on the applied voltage. Investigation of fronts propagating through a periodically modulated medium revealed the existence of a pinning range. Instead of a single parameter combination leading to a stationary front, Eq. (7), a whole range of parameters results in a stationary phase separation.[51]

Finally, we note that all controlled one-component models with cubic nonlinearity can be rescaled and cast in a particularly simple form. This can be seen as follows. Starting from the cubic reaction rate $R(u)$ expressed in terms of its three roots, Eq. (3), we introduce a rescaled concentration U, a new parameter α and rescaled space and time scales according to

$$U = \frac{u - u_1}{u_3 - u_1}, \quad t = \frac{T}{k(u_3 - u_1)^2}, \quad x = \frac{X}{(u_3 - u_1)}\sqrt{\frac{D}{k}}, \quad \alpha = \frac{u_2 - u_1}{u_3 - u_1}. \tag{20}$$

Now, the rescaled Schlögl model reads

$$\partial_T U = \partial_X^2 U + \tilde{R}(U) + \tilde{\mathcal{G}}(U)\tilde{f}(X, T). \tag{21}$$

The corresponding reaction function $\tilde{R}$, traveling front solution $\tilde{U}_c$, and velocity $\tilde{c}$ are obtained by substituting $u_1 \to 0$, $u_2 \to \alpha$, $u_3 \to 1$, $D \to 1$ and $k \to 1$ in Eq. (5), Eq. (6) and Eq. (3), respectively. This gives finally

$$\tilde{R}(U) = -U(U - \alpha)(U - 1), \tag{22}$$

$$\tilde{U}_c(x) = \frac{1}{1 + e^{\frac{x}{\sqrt{2}}}}, \tag{23}$$

$$\tilde{c} = \frac{1}{\sqrt{2}}(1 - 2\alpha). \tag{24}$$

Note that the coupling function is modified under rescaling, too,

$$\frac{1}{k(u_3 - u_1)^3}\mathcal{G}(u) = \frac{1}{k(u_3 - u_1)^3}\mathcal{G}((u_3 - u_1)U + u_1) = \tilde{\mathcal{G}}(U), \tag{25}$$

however, qualitative changes do not occur: a constant $\mathcal{G}$ stays constant under rescaling, a polynomial of degree n in U stays the same in $\tilde{\mathcal{G}}$, etc. The control signal $f(x, t) = f(X, T)$ is now expressed in the rescaled coordinates. We will use the rescaled Schlögl model (21) for all numerical simulations in the following chapters.

3. Nonlocal Control of Space-Time Patterns in the Schlögl Model

3.1. *Feedback control of the front propagation*

We consider feedback control of the RD system by adding either a distributed nonlocal feedback term (Eq. (26)) or a distributed time-delayed feedback (Eq. (27)):

$$\partial_t u = R(u) + \partial_x^2 u + \sigma \left[\int_{-\infty}^{+\infty} G(x')u(x - x')dx' - u(x) \right] \qquad (26)$$

$$\partial_t u = R(u) + \partial_x^2 u + \sigma \left[\int_0^{+\infty} G(t')u(t - t')dt' - u(t) \right] \qquad (27)$$

where $G(x)$ and $G(t)$ are normalized integral kernels whose integrals equal unity. Here we have chosen a form of the coupling term which is non-invasive for homogeneous or stationary states, respectively, i.e., it vanishes if $u(x,t)$ does not depend upon space (Eq. (26)) or time (Eq. (27)). If $G(t) = \delta(t-\tau)$, Eq. (27) reduces to Pyragas time-delayed feedback control.[52] Table 1 gives an overview of the investigated kernels.

The motivation for these distributed feedback forms comes from the elimination of one equation in the following two-variable reaction-diffusion system:

$$\partial_t u = R(u) - gw + \partial_x^2 u, \qquad (28)$$

$$\tau \partial_t w = hu - fw + D_w \partial_x^2 w, \qquad (29)$$

where all the terms are linear except the function $R(u)$, see Eq. (22). The two concentrations u and w correspond, respectively, to the activator and the inhibitor, and are linearly coupled by the terms $-gw$ and hu. The parameter D_w is the inhibitor diffusion coefficient ($D_w > 0$). The distributed nonlocal feedback term is obtained by the elimination of the second equation Eq. (29) in the limit when $\tau \to 0$.[7,53–57] The time-delayed feedback is obtained by eliminating the second equation in the limit when $D_w \to 0$.

3.2. Stability analysis of the homogeneous steady states

From Eq. (26) and Eq. (27), dispersion relations $\lambda(k)$ Eq. (30) and Eq. (31), respectively, are obtained by performing a linear stability analysis around the HSS solutions u_1, u_2, u_3 by setting $u(x,t) = u_{1,2,3} + \delta u$ with small $\delta u = e^{-ikx}e^{\lambda t}$:

$$\lambda = R'(u^*) - k^2 + \sigma(\sqrt{2\pi}\mathcal{F}\{G\}(k) - 1), \qquad (30)$$

$$\lambda = R'(u^*) - k^2 + \sigma(\mathcal{L}\{G\}(\lambda) - 1), \qquad (31)$$

where $\mathcal{F}\{G\}(k)$ and $\mathcal{L}\{G\}(\lambda)$ are the Fourier transform of the kernel $G(x)$ and the Laplace transform of the kernel $G(t)$, respectively.

From these dispersion relations, one can see that different kinds of instabilities may appear for suitable choice of σ: (i) nonlocal symmetric kernels may lead to a Turing instability ($\text{Im}(\lambda) = 0$, $\text{Re}(\lambda) > 0$ for finite wave number $k \neq 0$); (ii) nonlocal asymmetric kernels may lead to a traveling

Table 1. Kernels and their respective Fourier and Laplace transforms. (a) Nonlocal symmetric kernels. (b) Nonlocal asymmetric kernels. (c) Time-delayed kernels.

(a)	Name	$G(x)$	Fourier transform $\mathcal{F}\{G\}(k)$
	centered exponential	$\dfrac{1}{2a}e^{-\frac{|x|}{a}}$	$\dfrac{1}{\sqrt{2\pi}}\cdot\dfrac{1}{1+(ak)^2}$
	symmetric exponential[a]	$\dfrac{1}{4a}\left(e^{-\frac{|x-d|}{a}}+e^{-\frac{|x+d|}{a}}\right)$	$\dfrac{\cos dk}{\sqrt{2\pi}}\cdot\dfrac{1}{1+(ak)^2}$
	Gaussian	$\dfrac{1}{a\sqrt{2\pi}}e^{-\frac{x^2}{2a^2}}$	$\dfrac{1}{\sqrt{2\pi}}e^{\frac{-k^2a^2}{2}}$
	Mexican hat	$\dfrac{1}{a}\sqrt{\dfrac{2}{\pi}}\left(1-\dfrac{x^2}{2a^2}\right)e^{-\frac{x^2}{2a^2}}$	$\dfrac{2}{\sqrt{2\pi}}\left(1+\dfrac{a^2k^2}{2}\right)\right)e^{\frac{-a^2k^2}{2}}$
	centered rectangular	$a\Pi(ax)$	$\dfrac{1}{\sqrt{2\pi}}\cdot\text{sinc}\left(\dfrac{k}{2\pi a}\right)$
	symmetric rectangular[b]	$\dfrac{a}{2}\left(\Pi(a(x+d))+\Pi(a(x-d))\right)$	$\dfrac{\cos dk}{\sqrt{2\pi}}\cdot\text{sinc}\left(\dfrac{k}{2\pi a}\right)$

(b)	Name	$G(x)$	Fourier transform $\mathcal{F}\{G\}(k)$
	shifted exponential	$\dfrac{1}{2a}e^{-\frac{|x-d|}{a}}$	$\dfrac{e^{-idk}}{\sqrt{2\pi}}\cdot\dfrac{1}{1+(ak)^2}$
	shifted Gaussian	$\dfrac{1}{a\sqrt{2\pi}}e^{-\frac{(x-d)^2}{2a^2}}$	$\dfrac{e^{-idk}}{\sqrt{2\pi}}e^{\frac{-k^2a^2}{2}}$
	shifted rectangular	$a\Pi(a(x-d))$	$\dfrac{e^{-idk}}{\sqrt{2\pi}}\cdot\text{sinc}\left(\dfrac{k}{2\pi a}\right)$

(c)	Name	$G(t)$	Laplace transform $\mathcal{L}\{G\}(\lambda)$
	uniform delay[c]	$a\Pi(a(t-\tau))$	$\dfrac{2a}{\lambda}\sinh a\lambda/2e^{-\lambda\tau}$
	weak gamma delay	ae^{-at}	$\dfrac{a}{a+\lambda}$
	strong gamma delay	a^2te^{-at}	$\dfrac{a^2}{(a+\lambda)^2}$

[a]With $d/a \gg 1$, since $\int G(x)dx = 1 - \frac{1}{2}e^{-d/a}$.
[b]With $d - \dfrac{1}{2a} > 0$.
[c]With $\dfrac{a}{2} < \tau$.

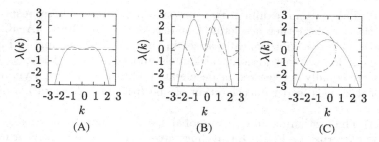

Fig. 2. Dispersion relation of the HSS u_1. (solid) Re $\lambda(k)$. (dashed) Im $\lambda(k)$. Obtained via (A) Eq. (30) with (A) the nonlocal symmetric kernel $G(x) = 1/(2a)\exp(|x|/a)$, (B) the nonlocal asymmetric kernel $G(x) = 1/(2a)\exp(|x-d|/a)$, and (C) Eq. (31) with the time-delayed kernel $G(t) = a\exp(-at)$. Parameters $\alpha = 0$, $a = 1$, $d = 2$ and $\sigma = -3$.

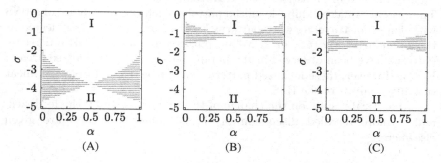

Fig. 3. Stability of the HSS u_1, u_3 in the (σ, α) parameter plan. Region I: both HSS are stable, i.e., Re $\lambda(k) < 0$ for both u_1, u_3. (dotted area): one of the HSS is unstable while the other is stable. Region II: both HSS are unstable, i.e., Re $\lambda(k) \geq 0$ for both u_1, u_3. Instabilities are: (A) Turing instability, (B) traveling waves instability and (C) Hopf instability. Results obtained via Eq. (30) with (A) the nonlocal symmetric kernel $G(x) = 1/(2a)\exp(|x|/a)$, (B) the nonlocal asymmetric kernel $G(x) = 1/(2a)\exp(|x-d|/a)$ and (C) Eq. (31) the time-delayed kernel $G(t) = a\exp(-at)$. Parameters $a = 1$, $d = 2$.

wave instability (Im$(\lambda) \neq 0$, Re$(\lambda) > 0$ for finite wave number $k \neq 0$); and (iii) time-delayed kernel may lead to a Hopf instability (Im$(\lambda) \neq 0$, Re$(\lambda) > 0$ for wave number $k = 0$). Figure 2 illustrates dispersion relations for these three different kinds of kernels. Furthermore, solving Re $\lambda(k) \geq 0$ leads to the phase diagrams of instabilities of the HSS shown in Fig. 3.

3.3. *Simulations*

Numerical simulations of Eqs. (26) and (27) have been performed. The behavior of the system can be classified into four types depending on the value of the coupling strength σ.

(i) The system exhibits two propagating fronts whose respective velocities can be either accelerated or decelerated, see Figs. 4A–4C and 5A, 5B. This behavior typically arises in the bistability regime in Fig. 3. Note that for symmetric nonlocal kernels and time-delayed kernels both fronts are equally accelerated or decelerated. On the contrary, asymmetric nonlocal kernels lead to acceleration of one front and deceleration of the other.

(ii) The system exhibits new global spatio-temporal patterns such as Turing patterns, see Fig. 4D, traveling wave patterns, see Fig. 4E or mixed wave patterns, see Fig. 5C. This behavior is characteristic of the wave regime in Fig. 3.

(iii) Between these two regimes, the system may exhibit transient patterns before asymptotically approaching a globally stable state. Figures 5D and 5E illustrates these transients for time-delayed feedback. The system can also exhibit localized patterns such as co-existence of a HSS with Turing or traveling wave patterns Figs. 4F and 4G, fronts reflected at the boundaries Fig. 5F, or traveling pulses Figs. 5G and 5H. All these patterns have been observed in the parameter regime of the dotted area of Fig. 3. However, these localized patterns cannot be derived from the linear stability analysis of the HSS.

Analytical results on the change of the front velocity by the feedback, and a more detailed discussion of the asymmetric kernels are given elsewhere.[7]

4. Position Control of Traveling Front Solutions to the Schlögl Model

Following[63] we consider the controlled Schlögl model of the form Eq. (8). We pursue a perturbative approach to the control problem and interpret the spatio-temporal control function $f(x,t)$ as a small term perturbing a stable traveling front solution $U_c(x)$. By multiple scale perturbation theory for small perturbation amplitude ϵ, the following equation of motion (EOM) for the position $\phi(t)$ of the perturbed front can be obtained,[34,35,58–62]

$$\dot{\phi} = c - \frac{\epsilon}{K_c} \int_{-\infty}^{\infty} dx\, e^{cx/D} U_c'(x) \mathcal{G}(U_c(x)) f(x + \phi, t), \qquad (32)$$

where $K_c = \int_{-\infty}^{\infty} dx\, e^{cx/D} \left(U_c'(x)\right)^2$ is constant and $\phi(t_0) = \phi_0$ denotes the initial condition. The derivation of Eq. (32) does not define a position of a front *a priori*, therefore we identify the point of steepest slope with the front position.

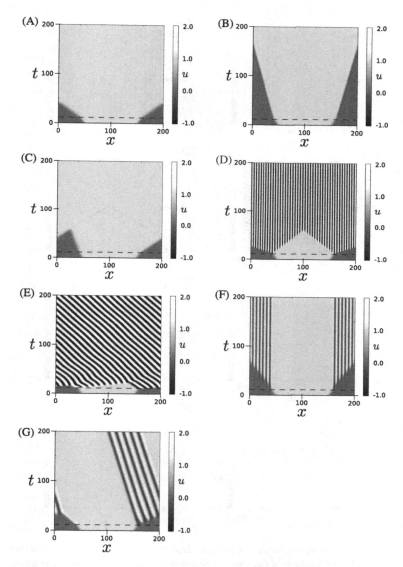

Fig. 4. Space-time patterns for distributed nonlocal feedback: (A) acceleration of the fronts; (B) deceleration of the fronts; (C) asymmetric acceleration; (D) Turing patterns; (E) traveling waves; (F) co-existence of Turing patterns and homogeneous state; (G) co-existence of traveling waves and homogeneous state. Results obtained via numerical simulation of Eq. (26) with $\alpha = 0$. (A)–(D) Gaussian kernel $G(x) = 1/(a\sqrt{2\pi}) \exp(-x^2/(2a^2))$, $a = 1$; (A) $\sigma = 5$; (B) $\sigma = -2$; (C) $\sigma = -3$; (D) $\sigma = -5$. (E)–(G) Shifted exponential kernel $G(x) = 1/(2a) \exp(-|x - d|/a)$, $a = 1$, $d = 5$; (F) $\sigma = 0.25$; (G) $\sigma = -1.0$; (H) $\sigma = -0.25$. Time scale (A)–(G) $0 \leq t \leq 200$, space scale (A)–(G) $0 \leq x \leq 200$.

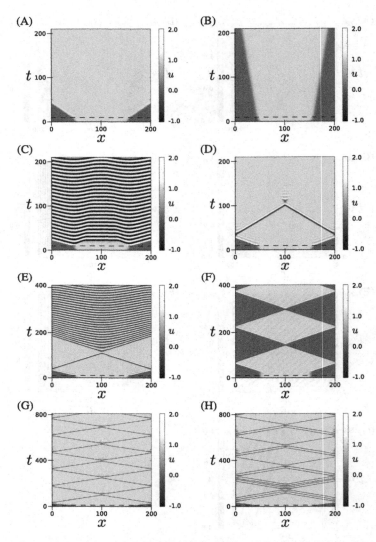

Fig. 5. Space-time patterns for distributed time-delayed feedback: (A) acceleration of the fronts; (B) deceleration of the fronts; (C) waves; (D) transient traveling pulse leading to HSS u_+^*; (E) transient traveling pulse leading to waves; (F) pairs of reflected fronts; (G) pairs of traveling pulses; (H) multiple fronts. Results obtained via numerical simulation of Eq. (27) with: (A)–(C) Strong gamma delay kernel $G(t) = a^2 t \exp(-at)$, $a = 1$, $\alpha = 0$, (A) $\sigma = -0.5$, (B) $\sigma = 2$, (C) $\sigma = -2$; (D)–(H) Weak gamma delay kernel $G(t) = a \exp(-at)$, (D) $a = 1$, $\alpha = 0$ $\sigma = -1.82$. (E) $a = 1$, $\alpha = 0.25$ $\sigma = -1.65$, (F) $a = 1$, $\alpha = 0.5$ $\sigma = -1.4$, (G) $a = 1$, $\alpha = 0$ $\sigma = -1.84$, (H) $a = 1$, $\alpha = 0.25$ $\sigma = -1.6$. Time scale (A)–(D) $0 \leq t \leq 210$; (E), (F) $0 \leq t \leq 410$; (G), (H) $0 \leq t \leq 810$. Space scale (A)–(G) $0 \leq x \leq 200$.

Here, we do not perceive Eq. (32) as an ordinary differential equation for the position $\phi(t)$ of the wave under the given perturbation f. Instead, Eq. (32) is viewed as an integral equation for the *control function* f.[63] The idea is to find a control which solely drives propagation in space according to an arbitrary prescribed *protocol of motion* $\phi(t)$. Simultaneously, we expect f to prevent large deformations in the uncontrolled wave profile $U_c(x)$. We assume that the wave moves unperturbed until reaching position ϕ_0 at time t_0, upon which the control is switched on.

A general solution of the integral equation (32) for the control f corresponding to the protocol of motion $\phi(t)$ is

$$f(x,t) = (c - \dot{\phi})\frac{K_c}{G_c}\, \mathcal{G}^{-1}(U_c(x - \phi))h(x - \phi), \qquad (33)$$

with constant $G_c = \int_{-\infty}^{\infty} dx\, e^{cx/D} U_c'(x) h(x)$. Here $\mathcal{G}^{-1}$ denotes the reciprocal of $\mathcal{G}$. The profile $\mathcal{G}^{-1}h$ of f is co-moving with the controlled wave and has constant amplitude. Equation (33) contains a so far undefined arbitrary function $h(x)$. A control proportional to the Goldstone mode U_c' shifts the front as a whole, simultaneously preventing large deformations of the wave profile. Therefore, in the following we choose $h(x) = U_c'(x)$, i.e.

$$f(x,t) = (c - \dot{\phi})\mathcal{G}^{-1}(U_c(x - \phi))U_c'(x - \phi). \qquad (34)$$

The constant K_c/G_c cancels out because $K_c = G_c$ for this choice. The control amplitude is entirely determined by the time dependent coefficient $c - \dot{\phi}$. Note that only the coupling function $\mathcal{G}$, the velocity c of the uncontrolled front, and the derivative of the traveling wave profile U_c enter the expression for the control signal. In particular, knowledge of the underlying reaction kinetics $R(u)$ or parameter values is not required. This makes the method useful for applications where model equations are only approximately known but the wave profile can be measured with sufficient accuracy.

For a first validity check, we assume a spatio-temporal control of the parameter α of the rescaled Schlögl model Eq. (23). This parameter can be seen as a measure for the excitation threshold of the Schlögl model: a localized perturbation nowhere exceeding this value cannot trigger a transition from the lower stationary stable state $u = 0$ to the upper stationary stable state $u = 1$. Substituting $\alpha \to \alpha + f(x,t)$ in Eq. (22) yields a coupling function

$$\mathcal{G}(u) = u(u - 1). \qquad (35)$$

The solution Eq. (34) for the control function immediately leads to

$$f(x,t) = (c - \dot{\phi}) \frac{U'_c(x - \phi)}{U_c(x - \phi)\left[U_c(x - \phi) - 1\right]}. \tag{36}$$

Plugging in the traveling front solution Eq. (23), we find a space-independent control function,

$$f(x,t) = \frac{1}{\sqrt{2}}(c - \dot{\phi}). \tag{37}$$

We interpret the result as follows. We set out to find a spatio-temporal control which changes the velocity of a traveling wave in a prescribed way and simultaneously preserves the uncontrolled front profile U_c. Because the front velocity $c = \frac{1}{\sqrt{2}}(1 - 2\alpha)$ depends linearly on α with coefficient $-\sqrt{2}$, we expect that an increase of parameter α by $\Delta\alpha = -\frac{1}{\sqrt{2}}\Delta c$ to a value $\tilde{\alpha} = \alpha + \Delta\alpha$ yields a front velocity $\tilde{c} = c + \Delta c$. Indeed, substituting $\phi(t) = \tilde{c}t = (c + \Delta c)t$ in Eq. (37) yields $f(x,t) = -\frac{\Delta c}{\sqrt{2}} = \Delta\alpha$. Because in the Schlögl model, the front profile does not depend on the parameter α, we achieved the desired goal of changing the front velocity while preserving the uncontrolled front profile for the case of constant protocol velocities $\dot{\phi} = \tilde{c} = \text{const}$. The result Eq. (37) can be seen as a generalization of fronts traveling at constant speed to arbitrary time dependent protocols of movement $\phi(t)$. As long as changes in the protocol velocity are slow, $|\ddot{\phi}| \ll 1$, and the maximum and minimum protocol velocities are sufficiently close to the velocity c of the uncontrolled front, we expect the control function (37) to result in a successful position control. Both assumption of slow changes in the front velocity and sufficiently small amplitude of perturbations are inherent in the multiple scale perturbation approach leading to the EOM (32) for perturbed traveling waves.

To demonstrate the performance of the proposed control approach, we consider an additive control $\mathcal{G}(u) = 1$ with a sinusoidal protocol

$$\phi(t) = B_0 + A\sin(2\pi t/T + B_1). \tag{38}$$

B_0 and B_1 are determined by $\phi(t_0) = \phi_0$, $\dot{\phi}(t_0) = c$ such that the protocol is smooth at the initial time t_0.

We carried out numerical simulations of the controlled Schlögl front with no-flux boundary conditions and used the uncontrolled wave profile $U_c(x)$ as initial condition. In Fig. 6, the obtained position and velocity over time data are compared with the prescribed protocol $\phi(t)$. The controlled front follows the prescribed protocol very closely. While front profile is

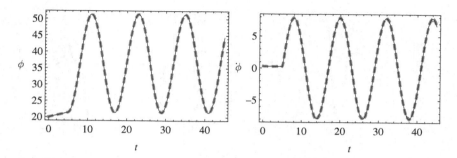

Fig. 6. Position control of a Schlögl front solution for an additive coupling function $\mathcal{G} = 1$. Position (left) and velocity over time data (right) obtained from numerical simulations of the controlled Schlögl front (red line) are in excellent agreement with the sinusoidal analytical protocol (black line). The parameter is $\alpha = 0.3$.

only slightly deformed by the control (not shown). Moreover, in many examples the results obtained with the control function (34) are in excellent agreement with the numerical solution of an optimal control problem where the traveling wave solution $U_c(x - \phi(t))$ shifted according to the protocol ϕ acts as the desired distribution enforced on the RD system.[63,64]

As a second example, we consider a coupling function motivated by the controlled liquid crystal light valve, Eq. (19), $\mathcal{G}(u) = b + du$. Using a smooth step function

$$\Theta_k(t) = (1 + \tanh(kt))/2 \qquad (39)$$

From which the discontinuous step function is recovered in the limit $\lim_{k \to \infty} \Theta_k(t)$, we define a smooth box function according to

$$B_k(t) = \Theta_k(1/2 - t) + \Theta_k(t + 1/2) - 1. \qquad (40)$$

Using Eqs. (39) and (40), we specify a protocol which drives the front to a certain position $\phi_0 + A_i$, stops it there for a given time interval w_i, moves it back to the initial position ϕ_0 and so on. The protocol consists of boxes of width w_i and amplitude A_i at times t_i,

$$\phi(t) = \phi_0 + \sum_i A_i B_k((t - t_i)/w_i). \qquad (41)$$

Under this control the front follows the prescribed protocol very closely, as the comparison with numerically obtained position and velocity over time data shows, see Fig. 7.

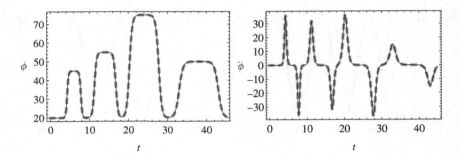

Fig. 7.   Comparison of the protocol with position (left) and velocity (right) over time data obtained by numerical simulation. The control moves a Schlögl front to a certain position, stops it there for a specified time interval, and moves it back (left). A coupling function $\mathcal{G}(u) = b + du$ with $b = 1$, $d = 1/2$ motivated by a liquid crystal light valve illuminated by a spatio-temporal light intensity is assumed. Black thin solid and red thick dashed lines display the analytical and numerical results, respectively. The parameter is $\alpha = 0.3$.

5. Conclusions

Using the Schlögl model as a paradigmatic example of a bistable RD system, we have discussed some physically feasible options of open- and closed-loop spatio-temporal control of RD systems. Control constraints arise from the physical meaning of the involved state and control variables representing concentrations, temperature, etc. Certainly, this aspect deserves more attention and further investigation with regard to the development of easy-to-realize experimental control schemes for RD systems.

We have shown, first, by open-loop control, how position and velocity of a traveling front can be precisely controlled in space and time without deforming the front profile substantially, and, second, how a variety of spatio-temporal patterns can be generated by feedback control. Our open-loop approach applies, if the front profile can be measured in the uncontrolled RD system with a precision that allows one to determine its spatial derivative with sufficient accuracy, Eq. (34). In addition, the control function $\mathcal{G}$ that depends on the intended control loop must be known. Remarkably, the often incompletely known nonlinear reaction kinetics is not required in order to set up the control. This makes the approach very promising if the underlying RD dynamics cannot be clarified in all details. Moreover, in two spatial dimensions, a generalized version of our approach allows for a precise and efficient control of the shape of wave patterns, too. In case of the closed-loop control special effects arise for asymmetric nonlocal kernels.[7]

References

1. Y. Kuramoto, *Chemical Oscillations, Waves and Turbulence* (Springer-Verlag, Berlin, 1984).
2. A. Mikhailov, *Foundations of Synergetics I: Distributed Active Systems* (Springer-Verlag, New York, 1990).
3. R. Kapral and K. Showalter (eds.), *Chemical Waves and Patterns* (Kluwer, Dordrecht, 1995).
4. T. Bánsági, V. K. Vanag and I. R. Epstein, *Science* **331**(6022), 1309–1312 (2011).
5. M. A. Dahlem, F. M. Schneider and E. Schöll, *Chaos* **18**, 026110 (2008).
6. F. M. Schneider, E. Schöll, and M. A. Dahlem, *Chaos* **19**, 015110 (2009).
7. J. Siebert, S. Alonso, M. Bär and E. Schöll, *Phys. Rev. E* **89**, 052909 (2014).
8. M. Kim, M. Bertram, M. Pollmann, A. von Oertzen, A. S. Mikhailov, H. H. Rotermund and G. Ertl, *Science* **292**, 1357 (2001).
9. Y. N. Kyrychko, K. B. Blyuss, S. J. Hogan and E. Schöll, *Chaos* **19**(4), 043126 (2009).
10. F. Tröltzsch (2010). *Optimal Control of Partial Differential Equations: Theory, Methods, and Applications*, Vol. 112 (American Mathematical Society, Providence).
11. A. Mikhailov and K. Showalter, *Phys. Rep.* **425**(2), 79–194 (2006).
12. Schimansky-Geier, B. Fiedler, J. Kurths and E. Schöll (eds.), *Analysis and Control of Complex Nonlinear Processes in Physics, Chemistry and Biology* (World Scientific, Singapore, 2007).
13. V. Vanag and I. Epstein, *Chaos* **18**(2), 026107–026107 (2008).
14. E. Schöll and H. G. Schuster (eds.), *Handbook of Chaos Control* (Wiley-VCH, Weinheim, 2008).
15. K. Agladze, V. Davydov and A. Mikhailov, *JETP Lett.* **45**(12), 767–770 (1987).
16. V. Zykov and H. Engel, *Physica D* **199**(1), 243–263 (2004).
17. J. Schlesner, V. Zykov, H. Brandtstädter, I. Gerdes and H. Engel, *New J. Phys.* **10**(1), 015003 (2008).
18. V. S. Zykov, G. Bordiougov, H. Brandtstädter, I. Gerdes and H. Engel, *Phys. Rev. Lett.* **92**, 018304 (2004).
19. E. Mihaliuk, T. Sakurai, F. Chirila and K. Showalter, *Phys. Rev. E* **65**(6); PART 1, 065602 (2002).
20. T. Sakurai, E. Mihaliuk, F. Chirila and K. Showalter, *Science* **296**, 2009–2012 (2002).
21. J. Schlesner, V. Zykov, H. Engel and E. Schöll, *Phys. Rev. E* **74**(4), 046215 (2006).
22. J. Wolff, A. G. Papathanasiou, H. H. Rotermund, G. Ertl, X. Li and I. G. Kevrekidis, *Phys. Rev. Lett.* **90**(1), 018302 (2003).
23. J. Wolff, A. G. Papathanasiou, I. G. Kevrekidis, H. H. Rotermund and G. Ertl, *Science* **294**(5540), 134–137 (2001).

24. J. Wolff (2002). *Lokale Kontrolle der Musterbildung bei der CO-Oxidation auf einer Pt (110)-Oberfläche*, Ph.D. thesis, FU Berlin, (2002).

25. J. Wolff, A. Papathanasiou, H. Rotermund, G. Ertl, M. Katsoulakis, X. Li, and I. Kevrekidis, *Phys. Rev. Lett.* **90**(14), 148301 (2003).

26. P. Kevrekidis, I. Kevrekidis, B. Malomed, H. Nistazakis and D. Frantzeskakis, *Phys. Scripta* **69**(6), 451 (2004).

27. H. Nistazakis, P. Kevrekidis, B. Malomed, D. Frantzeskakis and A. Bishop, *Phys. Rev. E* **66**(1), 015601 (2002).

28. B. Malomed, D. Frantzeskakis, H. Nistazakis, A. Yannacopoulos and P. Kevrekidis, *Phys. Lett. A* **295**(5), 267–272 (2002).

29. F. Schlögl, *Z. Phys. A* **253**(2), 147–161 (1972).

30. F. Schlögl, C. Escher and R. S. Berry, *Phys. Rev. A* **27**, 2698–2704 (1983).

31. E. Schöll, *Z. Phys. B* **62**, 245 (1986).

32. E. Schöll, *Nonlinear Spatio-Temporal Dynamics and Chaos in Semiconductors* (Cambridge University Press, Cambridge, 2001), Nonlinear Science Series, Vol. 10.

33. Y. B. Zel'dovich and D. A. Frank-Kamenetskii, *Dokl. Akad. Nauk SSSR* **19**, 693–798 (1938).

34. S. Alonso, J. Löber, M. Bär and H. Engel, *Europ. Phys. J. ST* **187**(1), 31–40 (2010).

35. J. Löber, M. Bär and H. Engel, *Phys. Rev. E* **86**, 066210 (2012).

36. D. Lebiedz and U. Brandt-Pollmann, *Phys. Rev. Lett.* **91**, 208301 (2003).

37. P. Érdi, *Mathematical Models of Chemical Reactions: Theory and Applications of Deterministic and Stochastic Models* (Manchester University Press, Manchester, 1989).

38. M. Korzukhin, Mathematical modeling of the kinetics of homogeneous chemical systems. II, in G. Frank (ed.), *Oscillatory Processes in Biological and Chemical Systems* (Nauk, Moscow, 1967a), pp. 231–242.

39. M. Korzukhin, Mathematical modeling of the kinetics of homogeneous chemical systems. III, in G. Frank (ed.), *Oscillatory Processes in Biological and Chemical Systems* (Nauk, Moscow, 1976b), pp. 242–251.

40. G. A. Papsin, A. Hanna and K. Showalter, *J. Phys. Chem.* **85**(17), 2575–2582 (1981).

41. A. Hanna, A. Saul and K. Showalter, *J. Am. Chem. Soc.* **104**(14), 3838–3844 (1982).

42. M. Bär, M. Falcke, C. Zülicke, H. Engel, M. Eiswirth and G. Ertl, *Surf. Sci.* **269**, 471–475 (1992).

43. S. Nettesheim, A. Von Oertzen, H. Rotermund and G. Ertl, *J. Chem. Phys.* **98**(12), 9977–9985 (1993).

44. E. Schöll, *Nonequilibrium Phase Transitions in Semiconductors* (Springer, Berlin, 1987).

45. F. Plenge, P. Rodin, E. Schöll, and K. Krischer, *Phys. Rev. E* **64**, 056229 (2001).

46. M. Meixner, P. Rodin, E. Schöll and A. Wacker, *Eur. Phys. J. B* **13**, 157 (2000).

47. M. Kehrt, P. Hövel, V. Flunkert, M. A. Dahlem, P. Rodin and E. Schöll, *Eur. Phys. J. B* **68**, 557–565 (2009).
48. E. Schöll, Pattern formation and time-delayed feedback control at the nanoscale, in G. Radons, B. Rumpf and H. G. Schuster (eds.), *Nonlinear Dynamics of Nanosystems* (Wiley-VCH, Weinheim, 2010), ISBN 978-3-527-40791-0, 325–367.
49. S. Residori, *Phys. Rep.* **416**(5–6), 201–272 (2005).
50. F. Haudin, R. Elías, R. Rojas, U. Bortolozzo, M. Clerc and S. Residori, *Phys. Rev. E* **81**(5), 056203 (2010).
51. F. Haudin, R. Elias, R. Rojas, U. Bortolozzo, M. Clerc and S. Residori, *Phys. Rev. Lett.* **103**(12), 128003 (2009).
52. K. Pyragas, *Phys. Lett. A* **170**, 421 (1992).
53. K. Kang, M. Shelley and H. Sompolinsky, *PNAS* **100**(5), 2848–2853 (2002).
54. E. M. Nicola, M. Or-Guil, W. Wolf and M. Bär, *Phys. Rev. E* **65**, 055101 (2002).
55. S.-I. Shima and Y. Kuramoto, *Phys. Rev. E* **69**(3), 036213 (2004).
56. P. Colet, M. A. Matias, L. Gelens and D. Gomila, *Phys. Rev. E* **89**, 012914 (2014).
57. L. Gelens, M. A. Matias, D. Gomila, T. Dorissen and P. Colet, *Phys. Rev. E* **89**, 012915 (2014).
58. L. Schimansky-Geier, A. S. Mikhailov and W. Ebeling, *Ann. Phys. (Leipzig)* **495**(4–5), 277–286 (1983).
59. A. Engel, *Phys. Lett. A* **113**(3), 139–142 (1985).
60. A. Engel and W. Ebeling, *Phys. Lett. A* **122**(1), 20–24 (1987).
61. A. Kulka, M. Bode and H. Purwins, *Phys. Lett. A* **203**(1), 33–39 (1995).
62. M. Bode, *Physica D* **106**(3), 270–286 (1997).
63. J. Löber and H. Engel, *Phys. Rev. Lett.* **112**, 148305 (2014).
64. R. Buchholz, H. Engel, E. Kammann and F. Tröltzsch, *Comput. Optim. Appl.* **56**, 153–185 (2013).

Chapter 12

FLOW-INDUCED CONTROL OF PATTERN FORMATION IN CHEMICAL SYSTEMS

Igal Berenstein* and Carsten Beta[†]

*Nonlinear Physical Chemistry Unit, Faculté des Sciences,
Université Libre de Bruxelles (ULB),
CP231, 1050 Brussels, Belgium

[†]Institute of Physics and Astronomy, University of Potsdam,
Karl-Liebknecht-Str. 24/25, 14476 Potsdam, Germany

Since Alan Turing's seminal paper in 1952, the study of spatio-temporal patterns that arise in systems of reacting and diffusing components has grown into an immense and vibrant realm of scientific research.[1,2] This field includes not only chemical systems but spans many areas of science as diverse as cell and developmental biology, ecology, geosciences, or semiconductor physics. For several decades research in this field has concentrated on the vast variety of patterns that can emerge in reaction-diffusion systems and on the underlying instabilities. In the 1990s, stimulated by the pioneering work of Ott, Grebogi and Yorke,[3] control of pattern formation arose as a new topical focus and gradually developed into an entire new field of research.[4] On the one hand, research interests concentrated on control and suppression of undesired dynamical states, in particular on control of chaos. On the other hand, the design and engineering of particular space-time patterns became a major focus in this field that motivates ongoing scientific effort until today.

Various approaches have been pursued to control pattern formation in reaction-diffusion systems, in particular, global external forcing and feedback. Here, a global system parameter is modulated according to an external rule (external forcing) or depending on the state of the system itself (feedback). Such strategies have been successfully implemented in different experimental systems like the Belousov–Zhabotinsky (BZ) reaction,[5,6] or the catalytic oxidation of CO on platinum.[7-9] Besides global control schemes, more versatile approaches have been explored. For example, non-uniform couplings can be introduced to generate structures of a desired size.[10] Localized pacemakers may suppress turbulence,[11] and combined global and local feedbacks can produce a rich variety of different dynamical states.[12]

While these control schemes act on a global or local quantity within the reaction-diffusion system, an entirely new class of control methods opens up if advection is included into the system. Here, we will review recent numerical work that explores the potential of a unidirectional flow to control pattern formation in a reaction-diffusion system. We will consider the effect of advection in systems with several different instabilities. First, we consider systems that show a Hopf bifurcation, i.e., systems that may undergo a transition from a stable steady state to periodic limit cycle oscillations upon a parameter change. Second, for a system with activator-inhibitor dynamics that displays a stable steady state, another source of instability occurs when the diffusivities of activator and inhibitor are different. In this case, a pattern develops that is periodic in space and stationary in time.[1] Third, in order to obtain a pattern that is periodic in both space and time, at least a three-variable system is needed in addition to different diffusion rates.[1] Besides systems that display these types of instabilities, we will also consider bistable systems under the influence of a unidirectional advective flow.

1. Hopf Bifurcation and Advection

Studies of pattern formation in oscillatory systems with unidirectional flow have been performed both experimentally and theoretically. Experiments have been carried out using the Belousov–Zhabotinsky (BZ) reaction, where a continuous flow stirred tank reactor feeds a tubular packed-bed reactor.[13,14] Also experiments with the CDIMA reaction have been performed.[15,16] Simulations have used different models such as the Oregonator,[17] the Brusselator,[18] and the Lengyel–Epstein model for the chlorine dioxide-iodine-malonic acid (CDIMA) reaction.[19] At the inflow boundary, the models use a Dirichlet boundary condition, at the outlet a no-flux condition.

In general, in these studies the formation of traveling waves was observed at low flow velocities and a stationary pattern known as *flow distributed oscillations* (FDOs) was observed at high flow velocities. To illustrate these findings, we have used a different model[20] consisting of one activator (x) and two inhibitors (y and z),

$$\partial x/\partial t = (1/\varepsilon)(x(1-x) - f_z z(x - q_z)/(x + q_z)$$
$$- f_y y(x - q_y)/(x + q_y)) + D_x \nabla^2 x + \phi \nabla x$$
$$\partial y/\partial t = a(x - y) + D_y \nabla^2 y + \phi \nabla y \tag{1}$$
$$\partial z/\partial t = \kappa x - (1 - a)z + D_z \nabla^2 z + \phi \nabla z.$$

For equal diffusivities ($D_x = D_y = D_z = 0.1$) and parameters $q_z = q_y = 0.01$, $f_z = 0.4$, $f_y = 1.4$, $a = 3/4$, $\varepsilon = 0.04$, $\kappa = 0.16$ we find for low values of ϕ, waves that travel with the flow, at intermediate flow speed waves traveling against the flow, and at high flow speed stationary structures (FDOs).[21] The wavelength of the FDOs is proportional to the flow speed as it is also seen for other systems.[17–19] The results are summarized in Fig. 1. It is worth noting that an equivalent system for reaction-diffusion-advection is a system in a growing domain,[22] but for these systems the structure is stationary with respect to the boundary.

In some cases, a spatially extended oscillatory system may develop spatio-temporal chaos even when the local dynamics are of the limit cycle type (diffusion induced turbulence). We recently found that this behavior

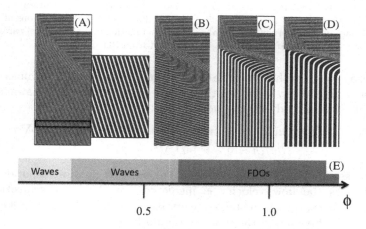

Fig. 1. Effect of advective flow on a system in the oscillatory regime. Space-time plots with size 80 space units (horizontal) and 400 time units (downwards). For the initial 100 time units the velocity ϕ is 0 and then is turned to 0.2 (A), 0.5 (B), 1.5 (C) and 2 (D). The square in (A), with 80 space units and 20 time units is expanded at the right. Light (dark) color represents high (low) concentration of the activator x. (E) Phase diagram for patterns at different velocities of the advective flow. Inlet is on the left hand side and the flow is from left to right. As parameters we used $D_x = D_y = D_z = 0.1$, $q_z = q_y = 0.01$, $f_z = 0.4$, $f_y = 1.4$, $a = 3/4$, $\varepsilon = 0.04$, $\kappa = 0.16$ (adapted from Ref. 21).

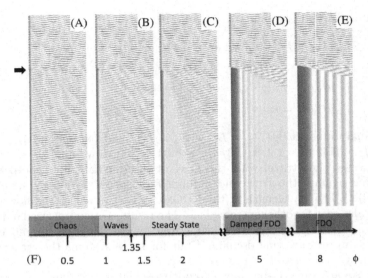

Fig. 2.   Control of chaos by advection. Space-time diagrams of u (high values represented by light color) for different values of the flow velocity ϕ. Space is displayed horizontally, time increased from top to bottom. Initially, we set $\phi = 0$. At the point in time indicated by the arrow on the left, the flow is initiated. We observe (A) defect-mediated turbulence at $\phi = 0.5$, (B) traveling waves at $\phi = 1.2$, (C) a uniform steady state at $\phi = 1.7$, (D) a damped periodic structure at $\phi = 5$, and (E) flow-distributed oscillations at $\phi = 8$. (F) Summary of the different dynamical regimes as a function of ϕ. The model parameters are $f = 2.1$, $q = 0.017$, and $D_u = D_v = 1$ (adapted from Ref. 23).

may also emerge in the two-variable Oregonator model at high values of the parameter f and low values of the parameter q.[23] The Oregonator model with diffusion and advection reads

$$\partial u / \partial t = (1/\varepsilon)(u(1-u) - fv(u-q)/(u+q)) + D_u \nabla^2 u + \phi \nabla u$$
$$\partial v / \partial t = u - v + D_v \nabla^2 v + \phi \nabla v. \tag{2}$$

We see in Fig. 2 that for small values of ϕ spatio-temporal chaos persists. As the flow velocity is increased, we observe a transition to traveling waves, then to a stable steady state, then to damped FDOs, and finally, at high values of the flow, we obtain FDOs.

2. Turing Instability and Advection

Theoretical work on this instability in the context of reaction-diffusion-advection systems has focused on the formation of stationary structures. In this case, the Turing instability obstructs the formation of stationary patterns.[24] To obtain a stationary structure, oscillatory behavior is needed besides the Turing instability, and the advection must be strong enough, so that the difference in diffusivities becomes negligible. To illustrate the effect of advection on a system with both the Hopf and the Turing instability, we performed simulations with the model system presented in Eq. (1) that was also used to produce Fig. 1. However, now the diffusion coefficients were chosen such that the activator x is diffusing slowly ($D_x = 0.1$) while both inhibitors diffuse fast ($D_y = D_z = 2$). Setting the other parameters as in Fig. 1, we see that even for the slightest flow the pattern drifts and only at high velocity we start seeing the beginning of formation of FDOs. At even higher velocities of the advective flow, FDOs are formed. The results are summarized in Fig. 3.

The only example for a system of this type that has been explored in experiments is the CDIMA reaction in growing domains.[25] At small velocities, stripes evolve that align with the flow, perpendicular to the boundary, whereas at high velocities the stripes orient parallel to the boundary (perpendicular to the flow). At intermediate values of the flow speed, the Turing stripes form at an angle to the boundary. It is worth

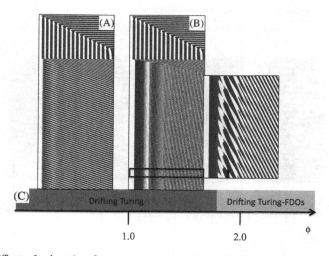

Fig. 3.   Effect of advective flow on a system with both Hopf and Turing instability. Space-time plots with size 80 space units (horizontal) and 400 time units (downwards). For the initial 100 time units the velocity ϕ is 0 and then is turned to 1.5 (A) and 3.0 (B). The square in (B), with 80 space units × 20 time units is expanded at the right. Light (dark) color represents high (low) concentration of the activator x. (C) Phase diagram for patterns at different velocities of the advective flow. As parameters we used $D_x = 0.1$, $D_y = D_z = 2$, $q_z = q_y = 0.01$, $f_z = 0.4$, $f_y = 1.4$, $a = 3/4$, $\varepsilon = 0.04$, $\kappa = 0.16$ (adapted from Ref. 21).

noting that the structures formed here are stationary, i.e., in the reference frame of the boundary (which is equivalent to a reaction-diffusion-advection system) these are moving patterns.

3. Wave Instability and Advection

For an appropriate choice of diffusion coefficients ($D_x = D_z = 0.1$ and $D_y = 2$) the three-variable system presented in Eq. (1) displays a wave instability.[20] Under these conditions, several different patterns can be produced by changing the parameters f_y and f_z. When taking advection into account, the variety of patterns increases.[20] A typical pattern that is obtained with a wave instability is standing waves. In the presence of advection, there is a transition from drifting-standing waves with out of phase oscillations between neighboring spots to synchronous (in-phase) standing waves, followed by a cascade of different patterns until finally FDOs are formed (see Fig. 4).

Also stationary patterns may emerge. Unlike the stationary patterns from a Turing instability, at low values of the advective flow, the pattern remains stationary and only adjusts its wavelength (see Fig. 5). A similar

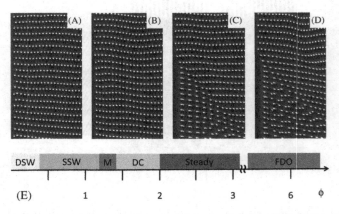

Fig. 4.   Effect of advection on standing waves in a system with wave instability. (A)–(D) Space-time plots with size 100 s. u. (horizontal) × 150 t. u. (downwards). Light (dark) color corresponds to high (low) concentrations of the activator x. The flow is turned on at 50 t. u. to $\phi = 0.1$ (A) 0.5 (B) 1.3 (C) and 1.5 (D). Drifting standing waves (DSW) are seen in (A), synchronous standing waves (SSW) in (B), mixed pattern (M) of drifting standing waves and synchronous standing waves in (C) and drifting clusters in (D). (E) Phase diagram, FDO stands for flow-distributed oscillations. As parameters we used $D_x = D_z = 0.1$, $D_y = 2$, $q_z = q_y = 0.01$, $f_z = 0.65$, $f_y = 1.9$, $a = 2/3$, $\varepsilon = 0.04$, $\kappa = 0.16$ (adapted from Ref. 20).

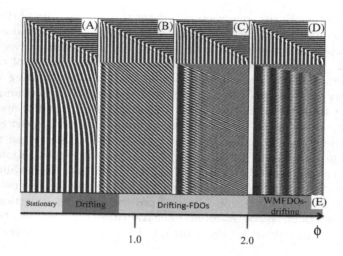

Fig. 5. Effect of advective flow on a system with wave instability that shows stationary structures. Space-time plots with size 80 space units (horizontal) and 400 time units (downwards). For the initial 100 time units the velocity ϕ is 0 and then is turned to 0.3 (A), 0.7 (B), 1.5 (C) and 2.5 (D). Light (dark) color represents high (low) concentration of the activator x. (E) Phase diagram for patterns at different velocities of the advective flow. WMFDOs stands for wave-modulated flow-distributed oscillations. As parameters we used $D_x = D_z = 0.1$, $D_y = 2$. $q_z = q_y = 0.01$, $f_z = 0.4$, $f_y = 1.4$, $a = 3/4$, $\varepsilon = 0.04$, $\kappa = 0.16$ (adapted from Ref. 21).

behavior is seen for waving stationary patterns, where low flow stabilizes the stationary structure.[20]

4. Bistability and Advection

How does a unidirectional flow influence pattern formation in systems that show co-existing stable solutions? We choose the Gray–Scott model as an example to pursue this question.[26] Depending on the choice of parameters, the Gray–Scott model can display bistability between two stable steady states, or between a steady state and a limit cycle. Furthermore, also spatio-temporal chaos is found in this model, resulting from irregular jumps between the steady state and the limit cycle.[27] The Gray–Scott model (with diffusion and advection) reads

$$\partial a/\partial t = 1 - a - \mu ab^2 + D_a \nabla^2 a - \phi \nabla a$$
$$\partial b/\partial t = b_0 - \sigma b + \mu ab^2 + D_b \nabla^2 b - \phi \nabla b. \tag{3}$$

In the case of two co-existing stable steady states, advection can induce the spreading of the less stable state into the more stable one. This is achieved if the velocity of advection is faster than the velocity of the trigger wave that propagates the more stable state into the less stable one.[28]

For co-existence of a limit cycle and a stable steady state, advection can affect the stability of the unstable steady state at the center of the limit cycle, similar to the system presented in Fig. 2. When starting from an oscillatory initial condition, we first observed disordered waves and defects emerging from the oscillatory state at low values of the flow speed. For increasing flow speeds, eventually the unstable focus in the center of the limit cycle is stabilized and gives way to damped FDOs, when the flow speed is increased even further, see Fig. 6. When the concentrations at the inflow boundary were chosen from the basin of attraction of the limit cycle, this was the final pattern reached for large flow speeds. However, when the inlet concentrations were chosen from the basin of attraction of the co-existing stable steady state, another transition occurred and this state was ultimately reached for high flow speeds, see Ref. 28 for more details.

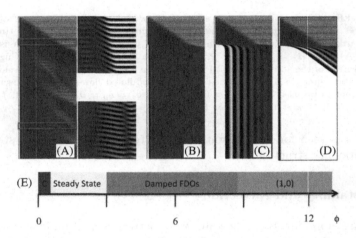

Fig. 6.   Effect of advective flow on a system with a stable node that co-exists with a stable limit cycle at $b_0 = 0$, $\mu = 155$ and $\sigma = 5$. Space-time plots are shown that have a size of 100 space units (horizontal) $\times$ 500 time units (vertical, time running from top to bottom). The velocity of the advective flow was 0 at the beginning. After 100 time units, it was turned to (A) 0.4, (B) 2, (C) 7, or (D) 9 space units per time unit. The gray scale represents the concentration of the reactant a, which spanned between 0 and 1. Dark corresponds to low and white to high values of a. A uniform initial condition of $(0.3, 0.6)$ was chosen. The Dirichlet boundary on the left was set to $(0,0)$ and a no-flux condition was imposed on the right. The two panels to the right of (A) show zooms of the regions marked by boxes on panel (A). (bottom) Phase diagram of patterns obtained as a function of the flow velocity ϕ. The steady state below $\phi = 3$ corresponds to the unstable focus in the center of the limit cycle, the steady state at $(1, 0)$ to the stable coexisting node (adapted from Ref. 28).

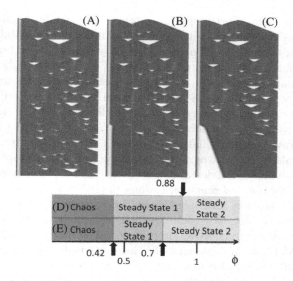

Fig. 7. Effect of advective flow on a Gray–Scott system in the chaotic regime with $b_0 = 0$, $\mu = 33.7$ and $\sigma = 2.8$. The space-time plots have a size of 100 space units (horizontal) × 750 time units (vertical, time running from top to bottom). At the beginning, the velocity of the advective flow is set to 0. After 500 time units, it is set to (A) 0.2, (B) 0.50, or (C) 0.80 space units per time unit. The gray scale represents the concentration of the reactant a, which spans between 0 and 1. Dark corresponds to low and white to high values of a. The Dirichlet boundary at the left is set to $(1,0)$, a no-flux condition is imposed on the right. In (D) and (E), phase diagrams summarize the patterns that were obtained for increasing advective velocities with a Dirichlet boundary of (D) $(1,0)$ and (E) $(0.1,0)$ (adapted from Ref. 28).

If parameters were chosen such that the systems displays spatio-temporal chaos, the advective flow will stabilize the unstable focus at moderate flow speeds, and furthermore trigger a second transition to the stable node $(1,0)$ for high flow speeds, see Fig. 7. Note that the transition from chaos to the stabilized focus occurs at a fixed value of the flow velocity, whereas the transition to the node at $(1,0)$ depends on the concentration values at the inflow boundary.

5. Conclusions and Outlook

We have explored a variety of spatio-temporal patterns that arise from the interaction between instabilities in reaction-diffusion systems and an advective flow. In particular, we have seen that advection can stabilize an otherwise unstable state, so that advection can be used as a simple control mechanism. Furthermore, systems that exhibit similar patterns show different responses to advection if the underlying instabilities are not the same. For example, with no-flux boundary conditions on both sides and no advection, the systems from Figs. 1, 3 and 5 all display bulk oscillations, and, as all parameters other than the diffusion coefficients are the same, the frequency of oscillations is also the same in all cases. Here, we can use advection to distinguish between the instabilities present in the system.[21]

So far, we have considered the case of equal advective flows for all species. However, in many natural situations, differential advection occurs, as for example in chromatography. For a system with a Hopf bifurcation, differential advection can destabilize a stable state, a phenomenon known as *differential flow induced chemical instability* (DIFICI). Indeed, experiments with the BZ reaction have confirmed the existence of this instability.[29] It will be interesting to explore, which new instabilities can arise due to coupling of different reaction-diffusion instabilities with differential advection. McGraw and Menzinger[30] have studied this case for a system with both differential diffusion and differential advection, but only for the Turing instability, not for the wave instability. They have also studied the effect of an oscillating boundary on pattern formation,[30] as well as the impact of a periodic modulation of the flow velocity,[31] where frequency locking is observed. Again, also in these cases, the wave instability has not been considered and will be the subject of future investigations.

Acknowledgments

The work presented here was mostly done while IB was at the University of Potsdam with a fellowship of the Alexander von Humboldt foundation.

References

1. A. M. Turing, *Philos. Trans. R. Soc. Lond. B. Biol. Sci.* **237**, 37–72 (1952).
2. M. C. Cross and P. C. Hohenberg, *Rev. Mod. Phys.* **65**, 851–1112 (1993).
3. E. Ott, C. Grebogi and J. A. Yorke, *Phys. Rev. Lett.* **64**, 1196–1199 (1990).
4. A. S. Mikhailov and K. Showalter, *Phys. Rep.* **425**, 79–194 (2006).
5. V. Petrov, Q. Ouyang and H. L. Swinney, *Nature* **388**, 655–657 (1997).
6. V. K. Vanag, L. Yang, M. Dolnik, A. M. Zhabotinsky and I. R. Epstein, *Nature*, **406**, 389–391 (2000).
7. M. Kim, M. Bertram, M. Pollmann, A. Von Oertzen, A. S. Mikhailov, H. H. Rotermund and G. Ertl, *Science.* **292**, 1357–1360 (2001).
8. C. Beta, M. Bertram, A. S. Mikhailov, H. H. Rotermund and G. Ertl, *Phys. Rev. E.* **67**, 046224 (2003).

9. M. Bertram, C. Beta, H. H. Rotermund and G. Ertl, *J. Phys. Chem. B.* **107**, 9610–9615 (2003).
10. C. Beta, M. G. Moula, A. S. Mikhailov, H. H. Rotermund and G. Ertl, *Phys. Rev. Lett.* **93**, 188302 (2004).
11. C. Punckt, M. Stich, C. Beta and H. H. Rotermund, *Phys. Rev. E.* **77**, 046222 (2008).
12. M. Stich and C. Beta, *Physica D* **239**, 1681–1691 (2010).
13. M. Kaern and M. Menzinger, *Phys. Rev. E.* **60**, R3471–R3474 (1999).
14. J. R. Bamford, R. Tóth, V. Gáspár and S. K. Scott, *Phys. Chem. Chem. Phys.* **4**, 1299–1306 (2002).
15. D. G. Míguez, G. G. Izús and A. P. Muñuzuri, *Phys. Rev. E.* **73**, 016207 (2006a).
16. D. G. Míguez, R. A. Satnoianu and A. P. Muñuzuri, *Phys. Rev. E.* **73**, 025201 (2006b).
17. J. R. Bamford, J. H. Merkin, S. K. Scott, R. Tóth and V. Gáspár, *Phys. Chem. Chem. Phys.* **3**, 1435–1438 (2001).
18. P. Andresén, M. Bache, E. Mosekilde, G. Dewel and P. Borckmanns, *Phys. Rev. E.* **60**, 297–301 (1999).
19. J. R. Bamford, S. Kalliadasis, J. H. Merkin and S. K. Scott, *Phys. Chem. Chem. Phys.* **2**, 4013–4021 (2000).
20. I. Berenstein, *Chaos*, **22**, 023112 (2012a).
21. I. Berenstein, *Chaos*, **22**, 043109 (2012b).
22. M. Kaern, R. A. Satnoianu, A. P. Muñuzuri and M. Menzinger, *Phys. Chem. Chem. Phys.* **4**, 1315–1319 (2002).
23. I. Berenstein and C. Beta, *J. Chem. Phys.* **135**, 164901 (2011).
24. A. Yochelis and M. Sheintuch, *Phys. Chem. Chem. Phys.* **12**, 3957 (2010).
25. D. G. Míguez, M. Dolnik, A. P. Muñuzuri and L. Kramer, *Phys. Rev. Lett.* **96**, 048304 (2006c).
26. P. Gray and S. K. Scott, *Chem. Eng. Sci.* **39**, 1087–1097 (1984).
27. R. Wackerbauer, and K. Showalter, *Phys. Rev. Lett.* **91**, 174103 (2003).
28. I. Berenstein and C. Beta, *Phys. Rev. E.* **86**, 056205 (2012).
29. R. Tóth, A. Papp, V. Gáspár, J. H. Merkin, S. K. Scott and A. F. Taylor, *Phys. Chem. Chem. Phys.* **3**, 957–964 (2001).
30. P. N. McGraw and M. Menzinger, *Phys. Rev. E.* **72**, 026210 (2005a).
31. P. N. McGraw and M. Menzinger, *Phys. Rev. E.* **72**, 027202 (2005b).

Chapter 13

DYNAMICS OF FILAMENTS
OF SCROLL WAVES

Vadim N. Biktashev

College of Engineering, Mathematics and Physical Sciences,
University of Exeter, UK

Irina V. Biktasheva

Department of Computer Science, University of Liverpool, UK

1. A Brief History and Motivation

One of notable events in the history of creation of the new science of "cybernetics" was Norbert Wiener's visit to Arturo Rosenblueth in Mexico, which resulted in their joint paper,[1] describing the first mathematical model of propagation of excitation pulses through a two-dimensional (2D) continuum, such as heart muscle. An important assertion of that theory was the possibility of the pulses to circulate around inexcitable obstacles, with important implications for understanding certain cardiac arrhythmias. Balakhovskii[2] realized that circulation of waves does not in fact require an obstacle, and the excitation wave may circulate "around itself", i.e., turning around its own refractory tail. Subsequently, such regimes of propagation became known as "reverberators", "rotors", "autowave vortices" and, mostly, "spiral waves" (see Fig. 1A). With the state of cardiac electrophysiology at the time, this concept remained a purely theoretical abstraction until the periodical chemical reaction discovered by Belousov[3] came to light and was further developed and investigated by Zhabotinsky[4,5] (Belousov–Zhabotinsky reaction, or just BZ) to fruition. The reaction was spontaneously oscillating, however in a non-stirred reactor, the fronts of the reaction oxidation stage were propagating similarly to electric pulses in cardiac muscle, and the spiral waves made by such propagation were observed.[6] The analogy with cardiac excitability was made even more

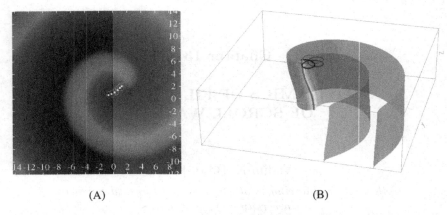

(A)   (B)

Fig. 1.   (A) Snapshot of spiral wave in the Barkley model (8) (u: red color component, v: blue color component), drifting in a stepwise inhomogeneity of paramer c (green color component), at base parameter values. The thin white line is the trace of the tip of the spiral, defined by $u(x, y, t) = u_*$, $v(x, y, t) = v_*$, in the course of a few preceding rotations. Yellow circles are positions of the centers calculated as period-averaged positions of the tip.[12] (B) Snapshot of scroll wave with buckled filament with negative tension in a thin layer of medium described by Barkley model (8).[13] Shown is the surface $u(x, y, z, t) = u_*$, colored depending on v; the green line is the instant filament defined as the locus of $u(x, y, z, t) = u_*$, $v(x, y, z, t) = v_*$. The gray scale line on the top face is the trace of the upper end of the filament. Parameter values: in (A), $a = 0.7$, $b = 0.01$, $c = 0.025$, $\delta c = 0.001$, $D_u = 1$, $D_v = 0$, $u_* = 0.5$, $v_* = 0.34$, box size 24×24; in (B), $a = 1.1$, $b = 0.19$, $c = 0.02$, $D_u = 1$, $D_v = 0.1$, $u_* = 0.5$, $v_* = 0.36$, box size $20 \times 20 \times 6.9$.

succinct by Winfree[7] who modified the BZ recipe to make the reaction dynamics excitable rather than oscillatory, and who was the first to present the BZ reaction and spiral waves to the Western audience. Being a physical model of cardiac tissue was arguably the most important use for the BZ reaction, motivating its study for all these years.

Since then, spiral waves have been observed in a wide variety of biological, chemical and physical systems, both artificial and in nature. We mention here just one example, the waves of cAMP signalling during the aggregation stage of social amoebae *Dictyostelium discoideum*.[8,9] There the spiral waves serve as organizing centers, as they provide signals to the individual amoebae where to crawl, to gather and merge into a multicellular organism and thus continue their peculiar lifecycle.

Unlike the Wiener and Rosenblueth's 2D theoretical abstraction, real excitable and oscillatory media, including BZ reaction and heart muscle, are three-dimensional (3D). Explicit experimental data on 3D extensions of spiral waves were first presented by Winfree[10] in his variant of BZ reaction. He also coined the term "scroll waves" (see Fig. 1B). While spiral waves rotate around their "cores", which can be considered point-like geometric objects, scroll waves rotate around "filaments", line-like geometric objects. Winfree also called them "organizing centres",[11] but in a sense different from what it means for the social amoebae, as there are no living bodies to receive the signals in the BZ reaction. Rather, the filaments organize the waves in the sense that the wave field in the whole volume follows what happens around the filament, and the rest of the wave field can be more or less predicted using Huygens principle, as shown by Wiener and Rosenblueth.

As spiral and scroll waves do not require any obstacles to rotate about, they can be located anywhere within the reactive medium. An inevitable, even if not immediately evident, consequence of that is that their position can change in time, i.e., they can drift, as illustrated by Fig. 1. Experimental and numerical studies of the drift have revealed that in many cases it is convenient to consider spiral waves as "particles", interacting with each other or reacting to external perturbations as localized, point-wise objects, the location being at the core around which the spiral rotates. The scroll waves in 3D have more degrees of freedom: their filaments cannot only move in space, but also change shape. The phase of rotation may vary along the filament, the feature known as "twist" of the scroll wave. Twist of a scroll wave and curvature of its filament are specifically three-dimensional factors of its dynamics.

It also turned out that not only it is *convenient to describe* motion of spirals and scrolls in terms of their cores and filaments, but it is *possible to predict* their motion in terms of cores and filaments coordinates *only* (particularly considering the phase as one of the coordinates). In this review, we aim to briefly discuss why this approach works and what sort of results it can produce. The literature on scroll dynamics is vast and the available space enforce us to restrict to selected examples in the narrow topic defined by the title of this chapter. We shall neglect plenty of other interesting results, including most of twist-related effects and everything related to competition and meander.

2. Wave-Particle Duality of Spiral Waves

The possibility to replace consideration of spiral waves by "particles", interacting with each other or reacting to external perturbations, is in a seeming contradiction with the wave-nature of these objects. The spiral waves "look" like nonlocalized objects, filling up all available space, but "behave" like localized objects. The mathematical nature of this paradox was brought to the forefront by Biktasheva and Biktashev[14] in terms of perturbative dynamics of spiral waves, i.e., drift of spirals in response to symmetry-breaking perturbations, such as spatial gradient of medium parameters or their resonant periodic modulation in time. Following Ref. 15, consider a reaction diffusion system

$$\partial_t \mathbf{u} = \mathbf{f(u)} + \mathbf{D}\nabla^2 \mathbf{u} + \epsilon \mathbf{h}, \tag{1}$$

where $\mathbf{u} = \mathbf{u}(\vec{r}, t)$, $\mathbf{f}, \mathbf{h} \in \mathbb{R}^\ell$, $\mathbf{D} \in \mathbb{R}^{\ell \times \ell}$, $\ell \geq 2$, $\vec{r} \in \mathbb{R}^2$, with perturbation $\epsilon \mathbf{h} = \epsilon \mathbf{h}(\vec{r}, t, \mathbf{u}, \nabla \mathbf{u}, \dots)$, $||\epsilon \mathbf{h}|| \ll 1$, and assume existence, at $\epsilon = 0$, of stationary rotating spiral solutions

$$\mathbf{u}(\vec{r}, t) = \mathbf{U}(\rho(\vec{r} - \vec{R}), \theta(\vec{r} - \vec{R}) + \omega t - \Phi) \tag{2}$$

where $\rho()$, $\theta()$ are polar coordinates centered at $\vec{R}$, constant ω is the angular velocity of spiral wave rotation, which up to the sign is uniquely defined by medium properties ($\mathbf{f}()$ and $\mathbf{D}$) (see, Sec. 4 below), and arbitrary constants $\vec{R} = (X, Y)$ and Φ are the location of the core of the spiral and its fiducial (initial) phase at $t = 0$. Then the first order perturbation theory in ϵ gives solutions close to (2) with $R = X + iY$ and Φ slowly varying according to motion equations

$$\dot{R} = \epsilon H_1(\vec{R}, \Phi, t), \qquad \dot{\Phi} = \epsilon H_0(\vec{R}, \Phi, t). \tag{3}$$

The velocities of spatial drift, H_1, and temporal/phase drift, H_0, are linear functionals of the perturbation,

$$H_n = \oint_{t-\pi/\omega}^{t+\pi/\omega} \frac{\omega \, dt'}{2\pi} \iint_{\mathbb{R}^2} d^2\vec{r} \, e^{in(\Phi - \omega t')} \langle \mathbf{W_n}(\rho, \theta + \omega t' - \Phi)), \mathbf{h} \rangle, \tag{4}$$

where $\mathbf{h}$ is evaluated at the unperturbed solution (2).

The kernels $\mathbf{W_{0,1}}$ of these functionals, which we call response functions (RFs), are critical eigenfunctions of the adjoint linearized problem. These eigenfunctions are dual to the eigenfunctions of the linearized problem produced by the generators of the Euclidean symmetry group, sometimes called Goldstone Modes (GMs). The "wave-particle" duality then reduces to the difference between these eigenfunctions. The GMs, constructed from spatial derivatives of the spiral wave solution, are nonlocalized. The RFs, however, are essentially localized, i.e., exponentially decay far from the core of the spiral. This is of course only possible because the linearized problem is not self-adjoint.

The spiral waves have localized response functions in both excitable and oscillatory media. Figure 2 shows response functions in the two-component

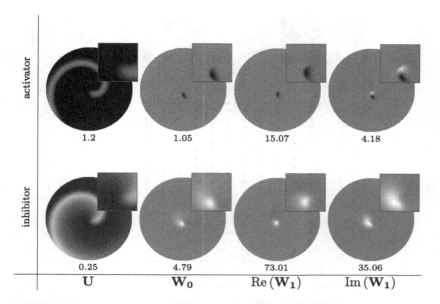

Fig. 2.   Density plots of the components of the spiral wave solution and its response functions in the Oregonator model (5) for $\varepsilon = 0.06$, $f = 1.75$, $q = 0.002$, $D_u = 1$, $D_v = 0.6$, the radius of the disk is 15 dimensionless units. In each plot, white corresponds to a value A and black corresponds to $-A$ where A is chosen individually for each plot, e.g., for the activator component of $\mathbf{U}$, $A = 1.2$. The grey periphery of the plots in columns 2–4 corresponds to 0. The central areas are also shown magnified in the small corner panels.

Oregonator[16] model of the BZ reaction,

$$\partial_t u = \frac{1}{\varepsilon}\left(u(1-u) - fv\frac{u-q}{u+q}\right) + D_u\nabla^2 u,$$
$$\partial_t v = u - v + D_v\nabla^2 v,$$

(5)

for a choice of parameters ε, f, q that gives excitable dynamics. Figure 3 shows response functions in the complex Ginzburg–Landau equation (CGLE),[18]

$$\partial_t w = w - (1-i\alpha)w|w|^2 + (1+i\beta)\nabla^2 w, \qquad w = u + iv \in \mathbb{C},$$

(6)

which is the "archetypical" oscillatory reaction-diffusion model, in the sense that it is a normal form of a reaction-diffusion system near a supercritical Hopf bifurcation in its reaction part. In CGLE, the response functions are localized for all sets of parameters where stable spiral wave solutions exist, with qualitative changes across critical lines in the parameter space.[17]

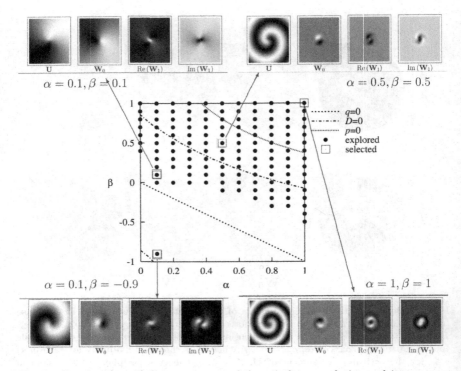

Fig. 3.   Density plots of the components of the spiral wave solution and its response functions in the complex Ginzburg–Landau equation (6), at various parameter values. In the legend: $p = 2(\alpha\beta - 1 + k^2(3 + 2\beta^2 - \alpha\beta))/(1 + \beta^2)$, $q = -4k((\alpha + \beta)(1 - k^2))(1 + \beta^2)$, $D = (p/3)^3 + (q/2)^2$, and $k = k(\alpha, \beta)$ is the asymptotic wave number of the spiral.[17]

Notice the non-monotonous behavior ("halo") in RFs close to the Eckhaus instability line (Fig. 3, bottom right inset). This can have phenomenological implication for the dynamics of spirals and scrolls, discussed later.

Apparently, the defining condition for the RFs' localization is the direction of the group velocity: a spiral wave will have localized RFs and behave as a localized object if and only if it is a source of waves, so that far from the core, the group velocity is directed outwards.[20,21]

Figure 4 reproduces two selected results illustrating how well the perturbation theory works, for two classical simplified excitable media models, the FitzHugh–Nagumo model[22]

$$\partial_t u = \frac{1}{\alpha}\left(u - \frac{u^3}{3} - v\right) + D_u \nabla^2 u,$$

$$\partial_t v = \alpha\left(u - \beta v + \gamma\right) + D_v \nabla^2 v, \qquad (7)$$

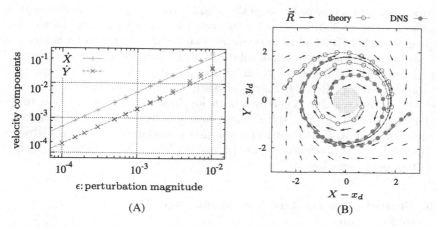

Fig. 4. Drift of spiral waves: asymptotic theory versus direct numerical simulations. (A) The velocity of the drift caused by the gradient of parameter a in the Barkley model (8), symbols for DNS and lines for the asymptotics, for base values of parameters $a = 0.7$, $b = 0.01$, $c = 0.025$, $D_u = 1$, $D_v = 0$. (B) The trajectories of the drift caused by a disk-shaped inhomogeneity in parameter γ in the FitzHugh–Nagumo model (7), at base values $\alpha = 0.3$, $\beta = 0.68$, $\gamma = 0.5$, $D_u = 1$, $D_v = 0$. The small arrows indicate drift velocities as predicted by the asymptotic theory, and the filled and open circles show the instantaneous centers of rotation of the spiral wave, measured with the interval of one period of rotation. See Refs. 12, 19 for detail.

and Barkley model[23]

$$\partial_t u = \frac{1}{c}u(1 - u)\left(u - \frac{v - b}{a}\right) + D_u\nabla^2 u,$$
$$\partial_t v = u - v + D_v\nabla^2 v. \tag{8}$$

The trajectories in Fig. 4B correspond to the case of the response function $\mathbf{W}_1$ with non-monotonic behavior (the "halos" in Fig. 3, bottom right, are an extreme case of such non-monotonic RFs). Here, small inhomogeneity attracts a spiral wave at larger distances and repels it at smaller distances. This alternating attraction/repulsion causes the spiral wave to "orbit" around at a certain stable distance, where the radial component of the "interaction force" vanishes.

There are important examples of spirals dynamics due to factors that are not small perturbations in the sense of (1), even though their action on the spirals is small. These include interaction of spiral waves with boundaries, and their interaction with each other (which may be considered as interaction of each spiral with the boundary between their "domains of influence"). These interactions are weak when the distance from the spiral core to the boundary or between the spiral cores is large. The mathematical aspects of particle-like behavior in such cases are less clear. In the few examples where analytical answers are known, this seems to be associated with the exponential growth of solutions of the non-homogeneous linearized problem with the free term given by the spatial gradient of the spiral wave solution, see e.g., Ref. 24. This is also stipulated by the outward direction of the group velocity. Therefore both localization properties seem to be equivalent. That is, if a spiral wave does not feel a weak inhomogeneity when far from it, it will not feel a non-flux boundary at the same distance. Although this equivalence is quite plausible physically, mathematically it is still an open question.

3. Perturbative Dynamics of Scrolls, and Tension of Filaments

The perturbative dynamics of spiral waves can be extended to scroll waves. In 3D, there are interesting phenomena even in absence of any symmetry breaking perturbations. Following Keener,[25] consider a generic reaction-diffusion system (1) in $\vec{r} \in \mathbb{R}^3$ at $\mathbf{h} \equiv 0$, and assume, as before, existence of stationarily rotating spiral solutions (2) in $\mathbb{R}^2$. A simple extension of spiral wave solution to the third spatial dimension is called a straight scroll wave. More generically, a scroll wave solution in $\mathbb{R}^3$ may be viewed as a solution of the form

$$\mathbf{u}(\vec{R} + \vec{N}\rho\cos\theta + \vec{B}\rho\sin\theta, t) = \mathbf{U}(\rho, \theta + \omega t - \Phi) + \mathcal{O}(\epsilon), \qquad (9)$$

where ϵ is now a formal small parameter measuring deformation of a scroll wave compared to the straight scroll, $\vec{R} = \vec{R}(\sigma, t)$ is the parametric equation of filament position at time t, $\Phi = \Phi(\sigma, t)$ is the rotational phase distribution, $\vec{N} = \vec{N}(\sigma, t)$ and $\vec{B} = \vec{B}(\sigma, t)$ are the unit principal normal and bi-normal vectors to the filament at point $\vec{R}(\sigma, t)$. Vectors $\vec{N}$ and $\vec{B}$ together with tangent vector $\vec{T}$ make a Frenet–Serret triple. In terms of the arclength differentiation operator $\mathcal{D}_s u(\sigma) = |\partial_\sigma \vec{R}|^{-1} \partial_\sigma u(\sigma)$, the tangent vector is $\vec{T} = \mathcal{D}_s \vec{R}$, the curvature κ and the normal unit vector $\vec{N}$ are defined by $\kappa\vec{N} = \mathcal{D}_s\vec{T}$, and the binormal vector $\vec{B} = \vec{T} \times \vec{N}$ completes the triad. The resulting filament's equation of motion, at small filament curvature, $\kappa = \mathcal{O}(\epsilon)$, and small twist, $\mathcal{D}_s\Phi = \mathcal{O}(\epsilon)$, can be written as[20]

$$\partial_t \vec{R} = \gamma_1 \mathcal{D}_s^2 \vec{R} + \gamma_2 [\mathcal{D}_s \vec{R} \times \mathcal{D}_s^2 \vec{R}] + \mathcal{O}(\epsilon^2), \qquad (10)$$

and it is decoupled from the evolution equation for the phase $\Phi(\sigma, t)$. Equation (10) is written in the assumption that parameter σ of the filament is chosen in such a way that a point with a fixed σ moves orthogonally to the filament (hence no component along $\vec{T}$).

The Frenet–Serret description is easy to understand but it has a significant disadvantage: it becomes degenerate at zero filament curvature, $\kappa = 0$. An alternative description, free from this disadvantage, is in terms of Fermi–Walker coordinates, corresponding to a Levi-Civita (torsion-free metric) connection along the filament (hereafter called Fermi coordinates for brevity). This changes the scroll's phase definition but does not affect (10) as it is decoupled.

The coefficients γ_1, γ_2 in (10) are calculated using the response functions $\mathbf{W}_1$ of the corresponding 2D spiral waves as

$$\gamma_1 + i\gamma_2 = -\frac{1}{2} \int_0^\infty \oint [\mathbf{W}_1(\rho,\theta)]^+ \mathbf{D} e^{-i\theta} \left(\partial_\rho - \frac{i}{\rho}\partial_\theta \right) \mathbf{U}(\rho,\theta) d\theta \, \rho \, d\rho. \quad (11)$$

Now let us consider the total length of the filament, defined at each t:

$$S(t) = \int ds = \int |\partial_\sigma \vec{R}| d\sigma. \quad (12)$$

Differentiation of (12), with account of (10) and using integration by parts, reveals that, neglecting boundary effects (absent for closed filaments and vanishing for smooth impermeable boundaries), the rate of change of the total length is

$$\frac{dS}{dt} = -\gamma_1 \int \kappa^2 ds + \mathcal{O}(\epsilon^2). \quad (13)$$

This implies that, within the applicability of the perturbation theory, unless the filament is straight, the total length of the filament decreases if $\gamma_1 > 0$ and increases if $\gamma_1 < 0$. Hence the coefficient γ_1 was called "filament tension" in Ref. 20.

Filament tension can be found via the asymptotic rate of shrinkage or expansion of large scroll rings with exact axial symmetry (see Fig. 5). This is only formally valid for scroll rings of sufficiently large radii in the unbounded space. Depending on the model, shrinkage of a scroll ring with positive filament tension may lead to its collapse, or may stop at some finite radius, while the ring continues to drift along its symmetry axis,[27,28] see Fig. 5. Possible theoretical explanations of this may involve higher-order corrections to (10) and/or interaction of different pieces of the scroll ring with each other, in the same way as 2D spirals interact; which if either of these is the dominant reason in any particular case is, as far as we are aware, currently an open question. Apart from simple scroll rings, there are more complicated structures with twisted, linked and/or knotted filaments that

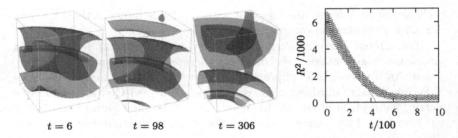

Fig. 5. Dynamics of a quarter of a scroll ring in FitzHugh–Nagumo model (7), at $\alpha = 0.30$, $\beta = 0.71$, $\gamma = 0.5$, $D_u = 1$, $D_v = 0$ (positive filament tension) in a box $33 \times 33 \times 40$ with periodic boundary conditions in the z direction. The right panel shows evolution of the square of the radius of the instant scroll filament, defined as the locus $u = 0$. Equation (10) predicts that the square of the period-average of the radius depends linearly on t, with the slope $-2\gamma_1$. The parameters for this example are taken from Ref. 26.

can be persistent in some models, see (Ref. 29, pp. 483–490) for the long story.

Filament tension depends on parameters of the medium, typically becoming negative in media with lower exitability, where spiral waves have larger cores.[30] This can be substantiated by the "kinematic" theory of excitation waves.[27] However, there are exceptions to the rule: e.g., Alonso and Panfilov[31] found an example where negative filament tension is observed at high excitability.

A simple but fundamental result is that when diffusivities of all the reagents are equal, $(\mathbf{D})_{ij} = D\delta_{ij}$, then $\gamma_2 = 0$ and $\gamma_1 = D > 0$.[32] A less trivial result is about filament tension in the CGLE (6), where it has been shown that $\gamma_2 = 0$, $\gamma_1 = 1 + \beta^2 > 0$, see e.g., Ref. 33 for discussion and references. So in both these cases the filament tension is guaranteed to be positive. We do not know of any generic results about negative filament tension.

A higher-order asymptotic equation for filament motion was obtained by Dierckx and Verschelde.[34] Unlike the leading order (10), it is coupled to the evolution of the phase:

$$\partial_t \phi = a_0 w^2 + b_0 \kappa^2 + d_0 \mathcal{D}_s w + \text{h.o.t.},$$

$$\partial_t \vec{R} = (\gamma_1 + a_1 w^2 + b_1 \kappa^2 + d_1 \mathcal{D}_s w)\mathcal{D}_s^2 \vec{R}$$

$$+ (\gamma_2 + a_2 w^2 + b_2 \kappa^2 + d_2 \mathcal{D}_s w)\mathcal{D}_s \vec{R} \times \mathcal{D}_s^2 \vec{R}$$

$$+ c_1 w \mathcal{D}_s^3 \vec{R} + c_2 w \mathcal{D}_s \vec{R} \times \mathcal{D}_s^3 \vec{R}$$

$$- e_1 \mathcal{D}_s^4 \vec{R} - e_2 \mathcal{D}_s \vec{R} \times \mathcal{D}_s^4 \vec{R} + \text{h.o.t.}, \qquad (14)$$

where ϕ is the scroll phase measured in the Fermi frame, and w is the corresponding twist, $w = \mathcal{D}_s \phi = \mathcal{D}_s \Phi - \tau$, where τ is the filament torsion, $\tau = \vec{B} \cdot \mathcal{D}_s \vec{N}$. The vector-function $\vec{R}(s, t)$ in (14) is the "virtual filament" which has to be defined more precisely than (9), and the coefficients are defined as integrals involving spiral wave response functions, similar to but more complicated than (11); see Ref. 34 for detail.

4. Scroll Wave Turbulence

A negative filament tension implies that the straight scroll of a sufficient lentgh should be unstable. It was recognized rather early, that unless restabilized by some mechanism beyond (10), such instability can lead to complex spatio-temporal behavior, possibly chaotic.[20,27] A particular interest for this complicated behavior was due to its possible relation to cardiac fibrillation, with which it would have a number of phenomenological features in common.[20,35,36] The predicted "scroll wave turbulence" was confirmed by numerical simulations, first in the FitzHugh–Nagumo model (7),[36] and then in Barkley model (8),[37] Oregonator model of the BZ reaction (5),[38] and Luo–Rudy model of heart ventricular tissue[39] to name a few; see Ref. 40 for a recent review.

Based on the generic results mentioned above, chances to observe scroll wave turbulence mediated by negative filament tension in a BZ reaction should be higher when some of the reagents are immobilized, since scalar diffusivity matrix implies positive filament tension; in the excitable rather than oscillatory regime, since tension in the CGLE is always positive; and preferably in the "lower excitability" case. The latter prediction concurs with Oregonator simulations by Alonso *et al.*[38]

We mentioned earlier that the angular velocity ω of the spiral rotation is uniquely determined by the properties of the medium. This is not strictly true. Typically, there is a discrete set of possible ω values, and in some cases there may be more than one of them. Winfree[41] identified a set of parameters in the FitzHugh–Nagumo model (7), which could, depending on initial conditions, support two alternative spiral wave solutions with different ω. Figure 7 shows a region in the parameter space with this property. Each of the alternative spiral waves is stable against small perturbations, but larger perturbations can convert one sort of spiral to the other. Foulkes *et al.*[26] investigated such conversions and some implications for the dynamics of scroll waves in 3D. One nontrivial effect observed there is shown in Fig. 8.

In various models, the straight scrolls may become unstable through mechanisms different from the negative tension in (10). Henry and Hakim[42] found that some parameter changes in Barkley model (8) can cause

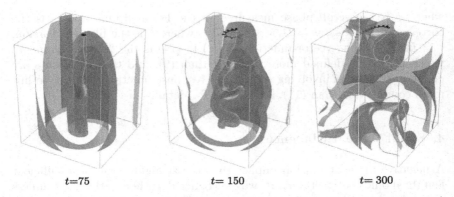

$t=75$ $t=150$ $t=300$

Fig. 6. Negative filament tension instability causes scroll wave turbulence. Barkley model (8) with parameters $a = 1.1$, $b = 0.19$, $c = 0.02$, $D_u = 1$, $D_v = 0$, box size $40 \times 40 \times 50$, instant filaments (green) defined as $u = v = 0.5$. Wavefronts are cut out by clipping planes halfway through the volume, to reveal the filaments.[13]

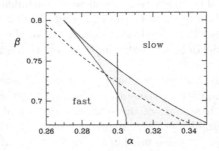

Fig. 7. Part of the parameter space in the FitzHugh-Nagumo model (7), giving alternative spiral wave solutions, at fixed $\gamma = 0.5$. The dashed line corresponds to zero tension. From Foulkes *et al.*[26]

finite-wavelength instability of a scroll, while the filament tension remains positive. A different type of finite-wavelength instability at positive filament tension was found in CGLE by Aranson and Bishop.[43] They interpreted it in terms of "self-acceleration" of spiral waves, at parameter values where the dynamics of spiral waves in 2D is not well described by the perturbative dynamics (3) and further corrections are required. Interestingly, these two finite-wavelength instabilities have different outcomes: in CGLE, this leads to turbulent-like behavior[43,44] visually similar to that shown in Fig. 6, whereas in Barkley model, it leads to re-stabilized "wrinkled" or "zig-zag" shape of filaments, as shown in Fig. 9.[13,42] Clearly, a mere linearized theory can not describe the outcome of any such instability and more detailed study would be required to make any predictions there.

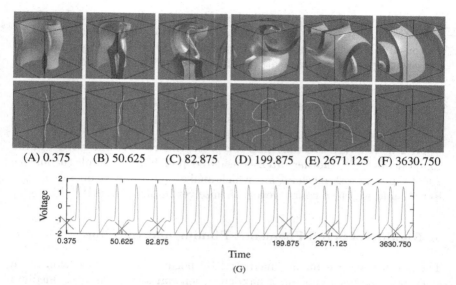

(A) 0.375   (B) 50.625   (C) 82.875   (D) 199.875   (E) 2671.125   (F) 3630.750

(G)

Fig. 8.   Evolution of a slow helical scroll with negative filament tension at $\alpha = 0.3$, $\beta = 0.71$, $\gamma = 0.5$, $D_u = 1$, $D_v = 0$, box size $50 \times 50 \times 50$. (A)–(F) (Top) isosurfaces of the u-field; (middle) filament ($u = v = 0$) only. (G) action potentials with blue crosses marking the times at which the snapshots were taken. The helix initially expands and turbulizes correspondingly to its negative tension, but then converts to its positive-tension alloform and contracts. From Foulkes et al.[26]

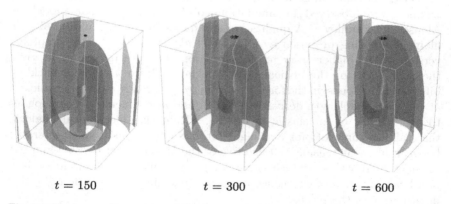

$t = 150$ $t = 300$ $t = 600$

Fig. 9.   Short-wave filament instability causes re-stabilized "wrinkled" filament. Barkley model with parameters $a = 0.66$, $b = 0.01$, $c = 0.025$, $D_u = 1$, $D_v = 0$, box size $40 \times 40 \times 50$, instant filaments (green) defined as $u = v = 0.5$. Wavefronts are cut out by clipping planes.[13]

Finally, complex turbulent-like behaviour of scroll waves may be related to spatial heterogeneity. Among these, we mention the cases described by Fenton and Karma in models with spatially varying anisotropy, aimed at representing characteristic features of cardiac ventricular muscles,

$$\partial_t \mathbf{u} = \mathbf{f}(\mathbf{u}) + \sum_{i,j=1}^{3} \partial_i (D^{i,j}(\vec{r})\partial_j)\mathbf{P}\mathbf{u}, \tag{15}$$

where $\mathbf{P} \in \mathbb{R}^{\ell \times \ell}$ is a constant matrix representing relative diffusivities of the components.[45,46] Another often considered possibility is scroll turbulence due to purely 2D mechanisms which make spiral waves unstable, e.g., Ref. 47 and which of course reveal themselves in 3D as well.[44,48–50]

5. Rigidity of Scroll Filaments: Pinning and Buckling

The scroll wave turbulence mediated by negative filament tension is an essentially 3D phenomenon: it happens when spiral waves in a 2D medium with the same parameters are perfectly stable. So such turbulence is not observed in quasi-two-dimensional domains, say in thin layers of reactive medium. A gradual increase of the reactive layer thickness reveals that between the stable quasi-two-dimensional behavior and the three-dimensional turbulence, there are intermediate regimes, one of which is the "buckled filament" illustrated in Fig. 1B for the Barkley model. Similar regimes were observed in simulations of Luo–Rudy cardiac model by Alonso and Panfilov.[39] A qualitative and quantitative explanation of such filament buckling proposed by Dierckx *et al.*[13] was based on a simplified version of the higher-order motion equations (14). In particular, it gives the expression for the critical layer thickness, above which the buckling bifurcation happens, in the form $L_* = \pi|e_1/\gamma_1|^{1/2}$, where e_1 is the coefficient at the fourth arclength derivative in (14), and in this sense it is analogous to the rigidity of an elastic beam. So, within this mechanical analogy, the stability threshold for a straight scroll is determined by the interplay between filament tension γ_1 (negative of the "mechanical stress") and the filament "rigidity" e_1, which is similar to the Euler's buckling instability of a beam under a stress, hence the term "buckling" used to describe this deformation of the scrolls.

Experimental evidence of scroll filament rigidity was demonstrated in BZ reaction, with filaments pinned to inexcitable inclusions,[51,52] see Fig. 10. In these experiments, the steady shapes of the filaments were determined, as well as their relaxation dynamics. The authors were able to describe the steady shapes using a variant of (14), with added phenomenological

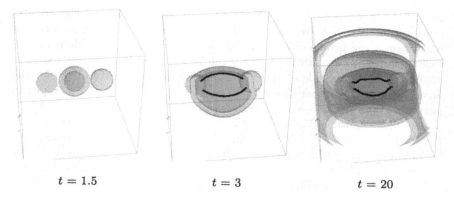

$$t = 1.5 \qquad\qquad t = 3 \qquad\qquad t = 20$$

Fig. 10. Schematic of the experiments done by Jiménez and Steinbock.[51] The upper half of a spherical wave is cut when it is passing through the beads thus forming two almost semi-circular filaments. The filaments become pinned to the beads and evolve to a stable configuration which depends on their tension, rigidity and interaction force between them. Oregonator model (5) with parameters $q = 0.002$, $f = 1.75$, $\varepsilon = 0.06$, $D_u = D_v = 0.5$, $u_* = 0.24$, $v_* = 0.075$, box size $50 \times 50 \times 50$. Wavefronts are cut out by clipping planes.

description of the interaction of filaments with each other. Fitting the theoretical curves to the experimentally found filament shapes allowed quantitative experimental measurement of the filament rigidity.[52]

6. Filament Statics, Geodesic Principle and Snell's Law

For cardiac applications, anisotropy and inhomogeneity are very important. We have already mentioned above the instability and scroll turbulence related to inhomogeneous anisotropy. The "opposite" of scroll turbulence is the situation when scroll dynamics converges to a stable equilibrium position. For positive filament tension, in a spatially uniform, isotropic medium and when interaction with boundaries or other filaments can be neglected, the answer is straightforward: a straight filament, stretching in any direction. Motivated by numerical simulations of scroll waves in models with anisotropic diffusion such as (15), and by experiments with re-entrant excitation waves in cardiac tissue, Wellner *et al.*[53] came up with a "geodesic hypothesis": that for given positions of the filament end points (say if the filament is anchored to inexcitable inclusions), the steady scroll filament shape will be a geodesic in a metric related to the diffusivity tensor, $g_{ij} = (D^{-1})_{ij}$. For instance, in a cardiac muscle, the preferrable orientation of a scroll filament would be along the fibers, i.e., direction of the maximal diffusivity of the transmembrane potential.

As ten Tusscher and Panfilov[54] noted, the metric defining the steady-state geometries of filaments, can be conveniently formulated in terms of excitation wave propagation time: given its end points, the filament will follow the *quickest* path connecting the end points. This hypothesis was confirmed by Verschelde *et al.*,[55] who generalized the equation (10) for the anisotropic media, under the assumption of $\det(D^{ij}) = $ const. An empirical generalization of this geodesic principle for non-uniform $\det(D^{ij})$, based on simulations using Barkley model, was suggested by Wellner *et al.*,[56] in the form $g_{ij} = \det(D^{ij})(D^{-1})_{ij}$. An interesting study has been performed by Zemlin *et al.*,[57] who investigated in numerical simulations the analogue of "Snell's law" for a filament crossing a boundary between media with different diffusivities, following from the geodesic principle, and the limits of its applicability.

References

1. N. Wiener and A. Rosenblueth, *Archivos del Instituto de Cardiologia de Mexico* **16**, 205–265 (1946).
2. I. S. Balakhovskii, *Biofizika* **10**, 1063–1067 (1965), in Russian.
3. B. P. Belousov, A periodic reaction and its mechanism, in *Sbornik referatov po radiatsionnoj medicine za 1958* (Medgiz, Moscow, 1959), pp. 145–147, in Russian. See also in R. J. Field and M. Burger (eds.), *Oscillations and traveling waves in chemical systems* (Wiley, New York, 1985).
4. A. M. Zhabotinsky, *Biofizika* **9**, 306–311 (1964b).
5. A. M. Zhabotinsky, *Proc. Ac. Sci. USSR* **157**, 392–95 (1964a).
6. A. M. Zhabotinsky and A. N. Zaikin, Spatial phenomena in the auto-oscillatory system, In E. E. Selkov, A. A. Zhabotinsky and S. E. Shnol (eds.), *Oscillatory Processes in Biological and Chemical Systems* (Nauka, Pushchino, 1971), p. 279.
7. A. T. Winfree, *Science* **175**, 634–636 (1972).
8. F. Alcantara and M. Monk, *J. Gen. Microbiol.* **85**, 321–334 (1974).
9. J. J. Tyson and J. D. Murray, *Development* **106**, 421–426 (1989).
10. A. T. Winfree, *Science* **181**, 937–939 (1973).
11. A. T. Winfree and S. H. Strogatz, Singular filaments organize chemical waves in three dimensions, *Physica*, **8D** 35–49 (1983), **9D**, 65–80, 333–345 (1983), **13D**, 221–233 (1984).
12. I. V. Biktasheva, D. Barkley, V. N. Biktashev and A. J. Foulkes, *Phys. Rev. E* **81**, 066202 (2010).
13. H. Dierckx, H. Verschelde, Ö. Selsil and V. N. Biktashev, *Phys. Rev. Lett.* **109**, 174102 (2012).
14. I. V. Biktasheva and V. N. Biktashev, *Phys. Rev. E* **67**, 026221 (2003).
15. V. N. Biktashev and A. V. Holden, *Chaos Solitons Fractals* **5**(3, 4), 575–622 (1995).

16. J. J. Tyson and P. C. Fife, *J. Chem. Phys.* **73**, 2224–2237 (1980).
17. I. V. Biktasheva and V. N. Biktashev, *J. Nonlin. Math. Phys.*, 8 Supp., 28–34 (2001).
18. Y. Kuramoto and T. Tsuzuki, *Prog. Theor. Phys.* **54**, 687–699 (1975).
19. V. N. Biktashev, D. Barkley and I. Biktasheva, *Phys. Rev. Lett.* **104**, 058302 (2010).
20. V. N. Biktashev, A. V. Holden and H. Zhang, *Philos. Trans. Roy. Soc. Lond. Ser. A* **347**, 611–630 (1994).
21. B. Sandstede and A. Scheel, *SIAM J. Appl. Dyn. Syst.* **3**(1), 1–68 (2004).
22. A. T. Winfree, *Chaos* **1**, 303–334 (1991b).
23. D. Barkley, *Physica D* **49**, 61–70 (1991).
24. V. N. Biktashev, Drift of a reverberator in an active medium due to interaction with boundaries, in A. V. Gaponov-Grekhov, M. I. Rabinovich, and J. Engelbrecht (eds.), *Nonlinear Waves II. Dynamics and Evolution* (Springer, Berlin, 1989), pp. 87–96.
25. J. P. Keener, *Physica D* **31**, 269–276 (1988).
26. A. J. Foulkes, D. Barkley, V. N. Biktashev and I. V. Biktasheva, *Chaos* **20**, 043136 (2010).
27. P. K. Brazhnik, V. A. Davydov, V. S. Zykov and A. S. Mikhailov, *Zh. Eksp. Teor. Fiz.* **93**, 1725–1736 (1987).
28. W. Skaggs, E. Lugosi and A. T. Winfree, *IEEE Trans. Circuits Syst.* **35**, 784–787 (1988).
29. A. T. Winfree, *The Geometry of Biological Time. Second edition* (Springer, USA, 2001).
30. A. V. Panfilov and A. N. Rudenko, *Physica* **28D** 215–218 (1987).
31. S. Alonso and A. V. Panfilov, *Phys. Rev. Lett.* **100**, 218101 (2008).
32. A. V. Panfilov, A. N. Rudenko and V. I. Krinsky, *Biofizika* **31**, 850–854 (1986).
33. I. S. Aranson and L. Kramer, *Rev. Mod. Phys.* **74**, 99–143 (2002).
34. H. Dierckx and H. Verschelde, *Phys. Rev. E* **88**, 062907 (2013).
35. A. T. Winfree, *Science* **266**, 1003–1006 (1994).
36. V. N. Biktashev, *Int. J. Bifurcation and Chaos* **8**, 677–684 (1998).
37. S. Alonso, R. Kahler, A. S. Mikhailov and F. Sagues, *Phys. Rev. E* **70**, 056201 (2004).
38. S. Alonso, F. Sagues and A. S. Mikhailov, *J. Phys. Chem. A* **110**, 12063–12071 (2006).
39. S. Alonso and A. V. Panfilov, *Chaos* **17**, 015102 (2007).
40. S. Alonso, M. Bär and A. V. Panfilov, *Bull. Math. Biol.* **75**, 1351–1376 (2013).
41. A. T. Winfree, *Physica D*, 49(1), 125–140 (1991a).
42. H. Henry and V. Hakim, *Phys. Rev. E* **65**, 046235 (2002).
43. I. S. Aranson and A. R. Bishop, *Phys. Rev. Lett.* **79**, 4174–4177 (1997).
44. J. C. Reid, H. Chaté and J. Davidsen, *EPL* **94**, 68003 (2011).
45. F. Fenton and A. Karma, *Phys. Rev. Lett.* **81**, 481–484 (1998a).
46. F. Fenton and A. Karma, *Chaos* **8**, 20–47 (1998b).
47. A. V. Panfilov and A. V. Holden, *Phys. Lett. A* **151**, 23–26 (1990).

48. R. H. Clayton and A. V. Holden, *Int. J. Bifurcation and Chaos* **13**, 3733–3745 (2003).
49. R. H. Clayton, E. A. Zhuchkova and A. V. Panfilov, *Prog. Biophys. Mol. Biol.* **90**, 378–398 (2006).
50. R. H. Clayton, *Chaos* **18**, 043127 (2008).
51. Z. A. Jiménez and O. Steinbock, *Phys. Rev. Lett.* **109**, 098301 (2012).
52. E. Nakouzi, Z. Jiménez, V. N. Biktashev and O. Steinbock, *Phys. Rev. E* **89**, 042901 (2014).
53. M. Wellner, O. Berenfeld, J. Jalife and A. M. Pertsov, *Proc. Natl. Acad. Sci. USA* **99**, 8015–8018 (2002).
54. K. H. W. J. ten Tusscher and A. V. Panfilov, *Phys. Rev. Lett.* **93**, 108106 (2004).
55. H. Verschelde, H. Dierckx and O. Bernus, *Phys. Rev. Lett.* **99**, 168104 (2007).
56. M. Wellner, C. W. Zemlin and A. M. Pertsov, *Phys. Rev. E* **82**, 036122 (2010).
57. C. W. Zemlin, F. Varghese, M. Wellner and A. M. Pertsov, Snell's law and the validity of minimal principle for scroll wave filaments (2014), to be published.

Chapter 14

UNUSUAL SYNCHRONIZATION PHENOMENA DURING ELECTRODISSOLUTION OF SILICON: THE ROLE OF NONLINEAR GLOBAL COUPLING

Lennart Schmidt[*,‡], Konrad Schönleber[*],
Vladimir García-Morales[*] and Katharina Krischer[*,†]

*Non-Equilibrium Chemical Physics - TU München,
James-Franck-Str. 1, D-85748 Garching, Germany
†krischer@ph.tum.de

‡Institute for Advanced Study - TU München,
Lichtenbergstr. 2a, D-85748 Garching, Germany

1. Introduction

Complex spatio-temporal pattern formation is an intriguing phenomenon often observed in nature. An example in biological systems is the spontaneous emergence of plane and spiral calcium waves in stimulated Xenopus oocytes.[1] Even human-body related phenomena are known, such as electrical turbulence in the heart muscle.[2,3] Spiral waves and their break-up could be observed in cardiac tissue and these phenomena could be explained from a solely dynamical point of view. Thus, for example, the spiral break-up is not triggered by inhomogeneities in the system, but rather arises via a dynamic instability. This immediately shows that there is a need for model systems to study the nonlinear dynamics in complex, pattern forming systems. Such a model system is the catalytic oxidation of carbon monoxide on a platinum surface in the UHV[4] exhibiting a variety of patterns, like spiral waves, pulses, solitons, target patterns and turbulence. Another model system is the Belousov–Zhabotinsky reaction, which became a prototypical system for the study of spiral dynamics, but gives also rise to standing, irregular and localized clusters under global feedback.[5,6]

In this chapter we present spatio-temporal dynamics found during the photoelectrodissolution of n-doped silicon and show that this experimental system is a convenient model system to study nonlinear dynamics with a conserved quantity: the mean-field of the two-dimensional oscillatory medium exhibits harmonic oscillations with constant amplitude and frequency. We model the experiments with an adapted version of the complex Ginzburg–Landau equation,[7,8] which is the appropriate normal form close to a supercritical Hopf bifurcation. In order to capture the mean-field oscillations, one has to introduce a nonlinear global coupling into the equation. As we will show in what follows, this nonlinear global coupling and the resulting conservation law are the origin of a wide variety of spatio-temporal patterns. Even the co-existence of regions exhibiting distinct dynamical behaviors is observed for many parameter values.

2. Experimental System

The experimental system under consideration is the potentiostatic photoelectrodissolution of n-doped silicon under high anodic voltage in the presence of a fluoride containing electrolyte. In this process, the silicon is first electrochemically oxidized and the oxide layer is subsequently etched away purely chemically by the fluoride in the electrolyte. As a result of this interplay, depending on the external voltage, a stable oxide layer may form.

The anodic oxidation follows either a tetravalent or a divalent mechanism, where in both cases the first stage is an electrochemical and the second stage a chemical process:[9,10]

$$
\begin{aligned}
\text{Si} + 4\text{H}_2\text{O} + \nu_{\text{VB}}h^+ &\rightarrow \text{Si(OH)}_4 + 4\text{H}^+ + (4 - \nu_{\text{VB}})e^- \\
\text{Si(OH)}_4 &\rightarrow \text{SiO}_2 + 2\text{H}_2\text{O} \quad \text{(tetravalent)} \\
\text{Si} + 2\text{H}_2\text{O} + \nu_{\text{VB}}h^+ &\rightarrow \text{Si(OH)}_2 + 2\text{H}^+ + (2 - \nu_{\text{VB}})e^- \\
\text{Si(OH)}_2 &\rightarrow \text{SiO}_2 + \text{H}_2 \quad \text{(divalent)}.
\end{aligned}
\tag{1}
$$

Here ν_{VB} is the amount of charge carriers stemming from the valence band of the silicon. As the divalent oxidation mechanism is accompanied by H_2 evolution the relative prevalence of both oxidation valences can be well distinguished. A significant contribution of the divalent mechanism is only found for relatively low anodic voltages[11] and the reaction valency ν in the parameter regime considered in our work is close to $\nu = 4$.[12] We will thus neglect the divalent oxidation pathway for the rest of this article.

The initial charge transfer step for the electrochemical oxidation is always the capture of a hole from the valence band of the silicon leading to $\nu_{VB} \geq 1$ in Eq. (1).[13] In n-type silicon these holes have to be photo-generated. Electron injections into the conduction band can lead to an overall current higher than the one induced by the photon flux incident on the surface, i.e. to external quantum efficiencies larger than one.[14] This current multiplication effect increases with decreasing illumination intensity and the limiting value of $\nu_{VB} = 1$ has been experimentally realized in literature.[11,14] While this trend is also present in our experiments, the values we typically find are in the range of $2 \leq \nu_{VB} \leq 4$.

The etching of the oxide is mainly due to the species HF and HF_2^- in dimolecular processes.[15] As the distribution of the fluorine to the species HF, HF_2^- and F^- is pH dependent and the rates for all dimolecular reaction pathways of the two etching species are different, the pH value as well as the total flourine concentration c_F determine the total etch rate. It is thus possible experimentally to vary the etch rate as well as the dominant etching pathway by the variation of the pH value of the solution and c_F. Especially the voltage dependence of the etch rate is expected to show some variation with the distribution of the flourine to the respective solvated species as HF_2^- and F^- are charged but HF is not.

A typical cyclic voltammogram of a silicon electrodissolution process together with a measure of the total mass of the corresponding oxide layer ξ is shown in Fig. 1A. Below ca. 0.2 V versus SHE (part I) the divalent oxidation mechanism dominates.[9,10,16] Increasing the voltage the tetravalent mechanism becomes dominant but no stable oxide layer forms as the etching process is faster than the oxidation (part II). At voltages higher than ca. 1.7 V versus SHE a stable oxide layer forms. This stable oxide layer leads to a decrease in the total current (part III). Starting at ca. 4 V versus SHE in the upward scan current oscillations can be seen. These oscillations become even more pronounced upon reversal of the scan

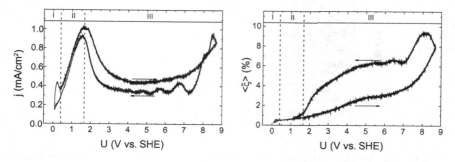

Fig. 1. Cyclic voltammogram (left) (20 mV/s) of a highly illuminated n-Si sample (pH = 1, c_F = 75 mM) and the corresponding, spatially averaged signal of the oxide layer mass ξ (right). The arrows indicate the scan direction.

direction and can also be seen in the mass of the oxide layer, as measured with an ellipsometric setup (see below) and depicted in Fig. 1(right). The difference between the current in the upward and the downward scan can be well understood by the corresponding difference in the mass of the oxide layer on the electrode surface inhibiting the current.

2.1. *Experimental setup*

To study the properties of the oxide layer during the potentiostatic photoelectrodissolution of n-doped silicon, we use a three electrode electrochemical cell equipped with an ellipsometric imaging system providing spatially resolved information on the oxide layer mass as shown in Fig. 2. The polarization of elliptically polarized light incident on the surface is changed upon reflection at the $Si|SiO_2|Electrolyte$ interface and this change is then converted into an intensity signal by a polarizer. Insertion of an imaging optic in the reflected light path gives an image of the electrode on a CCD chip. The system allows the *in situ* measurement of the optical path through the silicon oxide at a given area on the surface with a spatial resolution in the $10\,\mu m$ range. We call the intensity at each pixel relative to the detection limit of the CCD the ellipsometric intensity ξ.

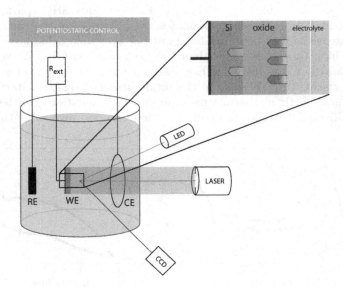

Fig. 2.   Sketch of the experimental setup showing the arrangement of the three electrodes and the external resistance together with the optical paths for the sample illumination (LASER) and the spatially resolved ellipsometric imaging (LED). A cross-section of the interface and the growth direction of the oxide is shown in the inset.

For an optimal contrast an angle of incidence on the sample surface close to the Brewster angle of the Si|Water system (ca. 70°) has to be chosen.

Silicon samples and electrolyte are prepared as described in Ref. 17 and the electrolyte is kept under an argon atmosphere throughout the experiments. To minimize parameter variations across the silicon surface the electrolyte solution is constantly stirred. Furthermore, the counter electrode is placed symmetrically opposite the silicon working electrode at a distance of several centimeters to minimize possible coupling effects.[18] Under these conditions perfectly uniform oscillations in ξ can be realized as shown in Fig. 3.

2.2. *Dynamics*

As early as 1958 it was established that the potentiostatic electrodissolution of p-doped silicon can proceed in an oscillatory fashion when the applied anodic bias is sufficient.[9] To stabilize the otherwise damped oscillations an external resistor connected in series with the working electrode has been found to be indispensible.[19] This external resistance R_{ext} links the potential drop across the silicon|oxide|electrolyte interface $\Delta\phi_{int}$ to the total current I passing through the surface.

$$\Delta\phi_{int} = U - R_{ext} \cdot I. \tag{2}$$

Thus, the external resistance introduces a coupling between all points at the electrode surface. This coupling is both global, as only the spatial average of the current is relevant, and linear. The behavior of the electrodissolution of n-doped silicon for sufficiently high illumination is identical to that of p-doped silicon.[20,21] This can be explained by the fact that the amount of holes in the valence band of the silicon is in this case always sufficient to maintain the current determined by the electrochemical parameters. In both cases the oscillations arise from a Hopf bifurcation occuring at a minimal, electrolyte specific threshold value of the external resistance[22] as shown in Fig. 4. It is clearly visible that with an increase of the external resistance first stable, sinusoidal oscillations with amplitudes increasing with the external resistance are found. This is the expected behavior close to a Hopf bifurcation. Upon further increase of R_{ext} the shape of the oscillations then gradually changes towards a more relaxational type. Above another electrolyte specific threshold value of R_{ext} the oscillations vanish abruptly and instead the system relaxes to a stable node. The upper boundary of the oscillatory regime with respect to the external resistance can be well understood by comparing the voltage drop across the interface in this case as shown in Fig. 4 to the CV scan shown in Fig. 1. The latter shows that

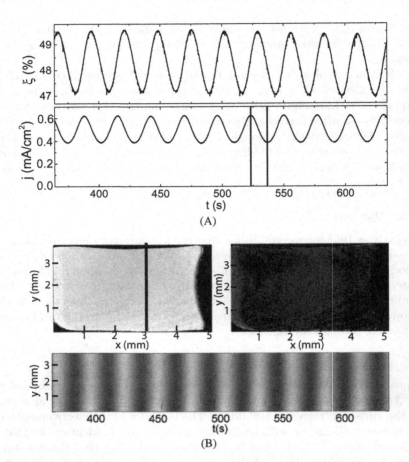

Fig. 3.   Uniform oscillation of ξ at a high illumination intensity and under constant potential ($c_F = 50\,\text{mM}$, pH $= 2.3$, $R_{\text{ext}}A = 2.7\,\text{k}\Omega\text{cm}^2$, $I_{\text{ill}} = 3.0\,\text{mW/cm}^2$, $U = 8.65$ V versus SHE). (A) Time series of the global quantities ξ and j; (B) Ellipsometric intensity distribution on the electrode for the two times indicated in (A) and the temporal evolution of a 1d cut along the line indicated in the left electrode picture. White indicates a relatively high and black a relatively low value of ξ.

the value of $\Delta\phi_{\text{int}}$ for R_{ext} above the oscillation boundary is in the voltage region where no stable oxide can form. The extent of the oscillatory regime is thus determined by a Hopf bifurcation at the low coupling limit and a cut-off caused by leaving the experimental parameters for a stable oxide in the high coupling limit.

A second coupling mechanism can be introduced by decreasing the illumination intensity, thus cutting off the total current by limiting the

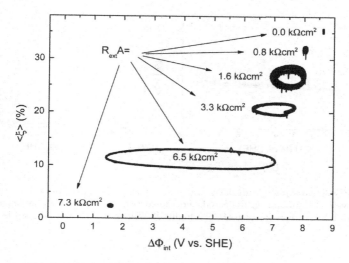

Fig. 4. Phase space plots of spatially averaged time series measured during the photoelectrodissolution of highly illuminated n-doped silicon (pH 1, $c_F = 75\,\text{mM}$, $U = 8.65$ V versus SHE) with varying R_{ext}.

generation of holes in the valence band of the n-doped silicon samples. A lower illumination intensity leads to a higher coupling strength. This coupling has a nonlinear characteristic and is at least partly global as the total current, i.e., the current averaged over all points on the surface, is again determining its strength. The cyclic voltammogram changes significantly when this coupling is introduced as shown in Fig. 5,[16] together with corresponding oxide mass changes. Comparing the illumination limited cyclic scans to the unlimited case, one notes that when the current reaches the illumination limit the oxide growth is initially suppressed. Only at significantly higher potentials does an oxide layer form. The potential shift for the oxide formation is illumination dependent and at too low illumination levels no oxide formation is found at all. In the cases where a stable oxide layer forms, again, both current and oxide layer mass show oscillations. The illumination limitation induced coupling is by itself sufficient to generate sustained oscillations as shown in the phase space plots in Fig. 6. Again the coupling strength is also a bifurcation parameter leading to stable foci below and stable oscillations above an electrolyte specific threshold value. At very high coupling strength the system relaxes to a stable node. In contrast to the case of the linear global coupling discussed above, however, the transition to the steady state is not abrupt and thus not of the same origin. It is linked to

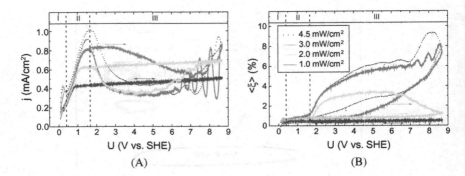

Fig. 5. Cyclic voltammogram (left) (20 mV/s) of an *n*-Si sample (pH = 1, c_F = 75 mM) at the levels of illumination indicated and the corresponding, spatially averaged signal of the oxide layer mass $\langle \xi \rangle$ (right). The highly illuminated case (dotted line) is identical to Fig 1.

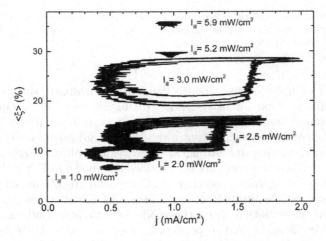

Fig. 6. Phase space plots of spatially averaged time series occurring during the photoelectrodissolution of *n*-doped silicon at various levels of the illumination intensity I_{ill} (pH 1, c_F = 75 mM) without an external resistance.

pattern formation and can thus only be understood considering the spatially extended system.

2.3. *Spatially extended system*

When regarding the spatially extended system an important difference between oscillations stabilized by the linear global coupling and oscillations stabilized by the nonlinear coupling becomes evident. In the former case the oscillations are always spatially uniform, while in the latter case they are often accompanied by pattern formation. This behavior was found in our group purely by chance as an n-doped silicon sample was erroneously used instead of a p-doped one. Unsurprisingly the behavior was quite unexpected and a current could only be seen when the lightproof box in which the experiment resided was opened. Under these conditions the patterns in ξ where first observed. Patterns are also found when the spatially resolved ellipsometric intensity recorded during the cyclic voltammograms shown in Fig. 1 and Fig. 5 is regarded. While in the highly illuminated case, i.e., at negligible nonlinear coupling, oxide growth and also the oscillations in ξ are spatially uniform, the oxide growth under restricted illumination proceeds along a growing wave front and the oscillations show spatial patterns in ξ. The nonlinear coupling is thus experimentally indispensible for pattern formation to occur. In experiments with a constant voltage, growing wave fronts are also present in the initial transients preceding the oscillations accompanied by pattern formation.

If both coupling mechanisms are combined, pattern formation is found as long as the linear global coupling is not too strong compared to the nonlinear coupling, leading to a typical parameter space as shown in Fig. 7.[21]

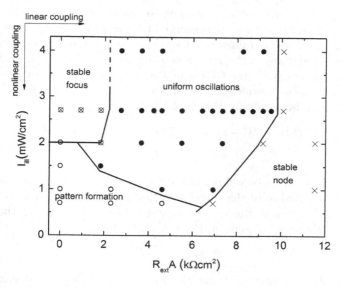

Fig. 7.   Dependence of the oscillation type on the strengths of the linear, global coupling $R_{\text{ext}} \cdot A$ and the nonlinear coupling, the restriction of I_{ill}.

In Sec. 4 we will present the patterns found in the experiments together with simulation results for a theoretical model system described in what follows.

3. Theoretical Modelling of Experiments

The experimental system presented in the preceding section can be modelled in a very general way. At high illumination intensities we observe homogeneous oscillations over the entire electrode surface. These oscillations originate in a Hopf bifurcation as described above (see Fig. 6). Thus, in order to model this system, the appropriate normal form to start with is the complex Ginzburg–Landau equation (CGLE)[7,8,23] for a complex order parameter $W(\mathbf{x}, t)$

$$\partial_t W = W + (1 + ic_1)\nabla^2 W - (1 + ic_2)|W|^2 W. \tag{3}$$

This equation describes all reaction-diffusion equations in the vicinity of a supercritics Hopf bifurcation. For a general derivation see Ref. 8. Equation (3) admits plane wave solutions of wave number Q

$$W_Q = a_Q \exp[i(\omega_Q t + Qx)], \tag{4}$$

with $|a_Q|^2 = 1 - Q^2$ and $\omega_Q = -c_2 + (c_2 - c_1)Q^2$.[8,23] A general solution is then given as a combination of these plane waves. This, in general, results in dynamics with an unpreserved homogeneous mode $W_0 = \langle W \rangle$. In contrast, for a huge parameter space the silicon system exhibits conserved harmonic oscillations in the averaged oxide-layer thickness. To achieve this in our model we extend the CGLE in a straightforward way by introducing a nonlinear global coupling into Eq. (3), leading to a modified complex Ginzburg–Landau equation (MCGLE)[24,25]

$$\partial_t W = W + (1 + ic_1)\nabla^2 W - (1 + ic_2)|W^2|W$$
$$- (1 + i\nu)\langle W \rangle + (1 + ic_2)\langle |W|^2 W \rangle. \tag{5}$$

Since we model a two-dimensional system, the complex order parameter $W(\mathbf{r}, t)$ is a function of the position vector $\mathbf{r} = (x, y)$ and time t. Angular brackets $\langle \cdots \rangle$ denote the spatial average. Now, when taking the spatial average of the whole equation, Eq. (5), one obtains

$$\partial_t \langle W \rangle = -i\nu \langle W \rangle, \tag{6}$$

which results in conserved harmonic oscillations of the spatial average,

$$\langle W \rangle = W_0 = \eta \exp[-i(\nu t + \phi_0)], \tag{7}$$

with amplitude η and frequency ν. ϕ_0 is an arbitrary initial phase. The essential dynamical properties of the silicon system are thus met with Eq. (5): oscillations arising through a Hopf bifurcation and the conserved harmonic mean-field oscillation. In Sec. 4 we show that this general ansatz indeed captures the pattern dynamics found in the experiments.

3.1. *Clusters*

The common notion of phase clusters describes a state, where the oscillatory medium separates into several parts. The oscillations in the different parts are phase shifted with respect to each other.[5,26–29] In the most simple case the clusters are arranged symmetrically and therefore the phase shifts in case of n clusters amount to $2\pi m/n$,[28–30] where $m = 1, 2, \ldots, n - 1$. Typically, in the case of cluster patterns the dynamics can be reduced to a phase model. However, there exist a second type of clusters, where essential variations in the amplitudes are present, called type II clusters.[24,25,31] We will see that this second type of cluster patterns naturally arises in our experiments and can be reproduced with the MCGLE.

But first of all we have to clarify, how clustering can occur and why it is possible in the MCGLE. Note that in the CGLE, Eq. (3), cluster patterns cannot emerge, since the dynamics are invariant under a phase shift $W \rightarrow e^{i\chi}W$, for arbitrary χ. For clustering to occur this symmetry has to be broken.[23] This becomes clear when considering two-phase clusters with a period of T_0 and phase balance (i.e., both clusters have the same size). Then, the dynamical equations have to stay invariant only when shifting the time t_0 to $t_0 + T_0/2$, but they are no longer invariant with respect to arbitrary shifts in time. In terms of the complex order parameter W this means that the dynamical equations are only invariant under the discrete transformation

$$W = \hat{W} \exp[i\omega_0 t] \rightarrow \hat{W} \exp[i(\omega_0(t + T_0/2))]. \tag{8}$$

With $\omega_0 = 2\pi/T_0$ the transformation reads

$$W \rightarrow e^{i\pi}W. \tag{9}$$

In general, for n clusters, one needs that the equations are invariant under the discrete symmetry[23]

$$W \rightarrow e^{i2\pi/n}W. \tag{10}$$

To account for this symmetry the proper extension of the CGLE is given by the term $\gamma_n W^{*n-1}$, describing also an external resonant forcing.[23,28,29,32–38] The asterisk denotes complex conjugation.

We will see that such a symmetry breaking term is intrinsically present in the MCGLE. Therefore, we write $W = W_0(1 + w)$ for the complex amplitude in Eq. (5). By exploiting the conservation law, Eq. (7), and the resulting fact that $\langle w \rangle = \langle w^* \rangle = 0$, one obtains again a CGLE, now for the inhomogeneity w, which reads[25]

$$\partial_t w = (\mu + i\beta)w + (1 + ic_1)\nabla^2 w$$
$$- (1 + ic_2)\eta^2(|w|^2 w + w^*) + C, \tag{11}$$

where

$$C = (1 + ic_2)\eta^2[\langle 2|w|^2 + w^2 \rangle - (2|w|^2 + w^2)] \tag{12}$$

and $\mu = 1 - 2\eta^2$, $\beta = \nu - 2c_2\eta^2$. Here the needed symmetry-breaking term $-(1 + ic_2)\eta^2 w^*$ occurs. But note that this term does not arise from the nonlinear global coupling. It would be present also when considering the CGLE, Eq. (3), without additional couplings. Crucial are the terms in C proportional to $|w|^2$ and w^2. As long as they are present, the equation is not symmetric with respect to the transformation $w \to e^{i\psi}w$ for any ψ. Here the nonlinear global coupling comes into play, as it renders a vanishing C possible, via the term proportional to $\langle 2|w|^2 + w^2 \rangle$. For this case, i.e., for $C = 0$, the occurrence of phase-balanced clusters is possible and the equation is symmetric with respect to the discrete symmetry $w \to e^{i\pi}w$. Note that this is impossible for a solely linear global coupling.

4. Results and Discussion

In the following sections we will demonstrate how well the dynamics of the oxide-layer thickness are captured with our very general ansatz in Eq. (5).

4.1. *Cluster patterns*

As we have clarified the theoretical basis for the emergence of cluster patterns in Sec. 3.1, we now turn towards the results of our experiments and simulations. In Fig. 8 we compare the cluster dynamics in the simulations of Eq. (5) with the experimental ones.[17] In Figs. 8A and 8C, two-dimensional snapshots of the simulations and the experiments, respectively, are shown. The spatio-temporal dynamics can be seen in one-dimensional cuts in Figs. 8B and 8D for the simulations and the experiments, respectively. They show that the homogeneous oscillation is modulated by two-phase clusters. Therefore, it is clear, that the phase shift between two regions is not given by π.

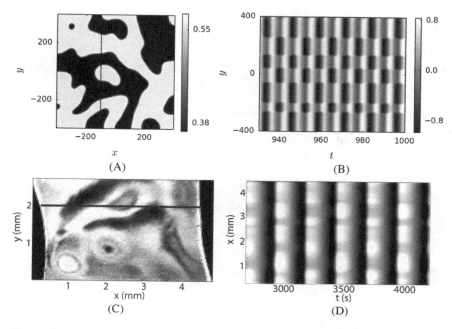

Fig. 8.   Two-phase clusters in theory (A), (B) and experiment (C), (D). (A) Snapshot of the two-dimensional oscillatory medium in the theory. Shown is the real part of W. (B) Spatio-temporal dynamics in an one-dimensional cut versus time in the theory. (C), (D) The same as (A), (B) now for the experimental results. The simulation captures the experimental dynamics very well. Note that the colorbars are different for each sub-figure. Parameters read: $c_1 = 0.2$, $c_2 = -0.58$, $\nu = 1.0$, $\eta = 0.66$ (simulation) and $c_F = 35\,\text{mM}$, pH $= 1$, $R_{ext} \cdot A = 9.1\,\text{k}\Omega\text{cm}^2$, $I_{ill} = 0.7\,\text{mW/cm}^2$ (experiment).

To analyze time series of this we perform a Fourier transformation in time at every point $\mathbf{r}$ of the ellipsometric signal and of the real part of $W(\mathbf{r},t)$ for the experiments and simulations, respectively.[24] We spatially average the resulting amplitudes $|a(\mathbf{r},\omega)|^2$ to obtain the cumulative power spectrum $S(\omega) = \langle |a(\mathbf{r},\omega)|^2 \rangle$. Results are shown in Fig. 9.

Two major peaks occur in both cumulative power spectra in Figs. 9A (theory) and 9D (experiment), one at the frequency ν of the mean-field oscillation. The other one describes the frequency of the clusters. This becomes clear when considering the Fourier amplitudes, corresponding to these peaks, in the complex plane: At $\omega = \nu$ (Figs. 9B and 9E) all local oscillators form a bunch, while at $\omega \approx \nu/2$ (Figs. 9C and 9F) the oscillators arrange into two clusters, located at the endpoints of the bar visible. Due to the diffusive coupling, the clusters are connected by an interfacial region, leading to the intermediate oscillators of the bar. The fact that

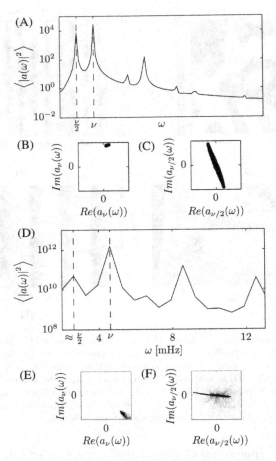

Fig. 9. (A), (D) Cumulative power spectra for simulations and experiments, respectively. The two major peaks at ν and approximately $\nu/2$ are indicated. The arrangement of the local Fourier amplitudes in the complex plane corresponding to these peaks are depicted in (B) and (C) for the theory and (E) and (F) for the experiments, respectively. The whole two-dimensional system is considered, which leads to the scattered oscillators in the experimental result in (F).

the connection of the two clusters crosses the zero point implies that the borders between them are Ising-type walls. As in this picture the phase shift between the two clusters is given by π, we conclude that at this frequency the clustering takes place. In the experiments in most cases the cluster frequency is given by approximately $\nu/2$. This leads to the conclusion that the clusters arise via a period-doubling bifurcation. Contrarily, in the theory the cluster frequency can be tuned continuously. For better comparison, we chose the parameter values such that the frequency also amounts to $\nu/2$.

4.2. *Subclustering*

A symmetry-breaking type of clustering also occurs in our simulations and experiments. The system again separates into two regions as in the case of the two-phase clusters, but now one region is homogeneous, while the other one exhibits two-phase clusters as a substructure.[17] Such states were also observed in Refs. 39, 40. The results are depicted in Fig. 10.

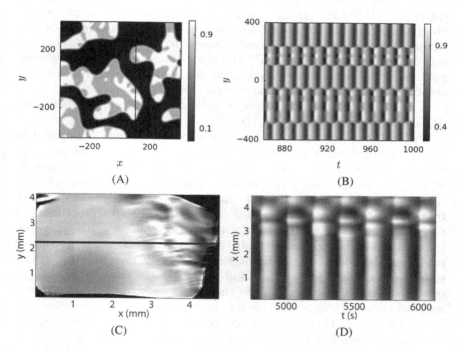

Fig. 10. Sub-clustering in theory (A), (B) and experiment (C), (D). Again snapshots (A), (C) and one-dimensional cuts (B), (D) are shown. The system splits into two regions, one region being homogeneous and one exhibiting two-phase clusters as a sub-structure. Note that in (B) for better visibility $|W(y, t)|$ is shown. Parameters read: $c_1 = 0.2$, $c_2 = -0.67$, $\nu = 0.1$, $\eta = 0.66$ (simulation) and $c_F = 35\,\mathrm{mM}$, pH $= 1$, $R_{ext} \cdot A = 7.6\,\mathrm{k\Omega cm^2}$, $I_{ill} = 0.5\,\mathrm{mW/cm^2}$ (experiment).

In the simulations the two-phase subclusters oscillate at half the frequency of the main clusters. This indicates that the subclustering is connected to a period doubling bifurcation.

4.3. *Chimera states*

In the preceding section we demonstrated that the symmetry of the two-phase cluster pattern can be broken, resulting in a substructure in one of the two domains. This symmetry-breaking can be even more dramatic: one of the two domains does not exhibit a coherent substructure, it rather displays turbulent dynamics. Thus the whole system separates into two regions, one being synchronized, while the other one displays incoherent and chaotic oscillations. This co-existence of synchrony and asynchrony was termed a chimera state[41] and many theoretical investigations deal with this topic.[17,41–50] Chimera states might be of importance for some peculiar observations in different disciplines, such as the unihemispheric sleep of animals,[51,52] the need for synchronized bumps in otherwise chaotic neuronal networks for signal propagation[53] and the existence of turbulent-laminar patterns in a Couette flow.[54] They could also be realized experimentally in chemical, optical, mechanical and electrochemical systems.[17,40,55–57]

In Fig. 11 we present the chimera states found in the simulations (A), (B) and in the experiments (C), (D).[17]

In both the experiments and the simulations nothing is imposed to induce this symmetry-breaking. The experimental conditions are kept uniform over the entire electrode. Furthermore, these patterns form spontaneously, i.e., no specially prepared initial conditions are required to obtain them. We could show in Ref. 17 that these peculiar dynamics can be traced back to the nonlinear global coupling. In order to proof this, we performed simulations of Eq. (5) without the diffusive coupling, which means that we dealt with an ensemble of Stuart–Landau oscillators coupled solely via the nonlinear global coupling present in Eq. (5). Also in this system the chimera state arises and has the same features as the one presented here. Thus, contrarily to the convincement in literature that a nonlocal coupling is indispensable for the emergence of chimera states, the chimera state found in our simulations forms under a solely global coupling. The diffusive coupling leads to the spatial arrangement into synchronized and incoherent regions. In the experiments a nonlocal contribution of the nonlinear coupling could not be ruled out yet. However, the striking similarity between simulations and experiments strongly corroborates the notion that the nonlinear coupling acts essentially globally.

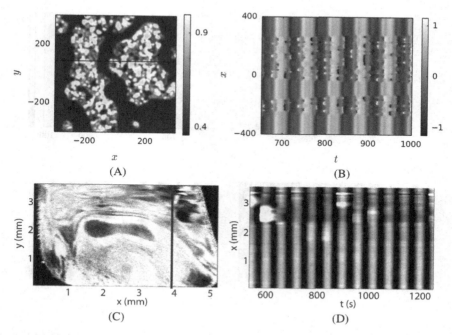

Fig. 11.   Chimera states in theory (A), (B) and experiment (C), (D). Snapshots (A), (C) and one-dimensional cuts (B), (D) are shown. In both, the simulation and the experimental pattern, the synchronized and turbulent regions can be clearly distinguished. Parameters read: $c_1 = 0.2$, $c_2 = -0.58$, $\nu = 0.1$, $\eta = 0.66$ (simulation) and $c_F = 50$ mM, pH $= 3$, $R_{ext} \cdot A = 4.5$ kΩcm^2, $I_{ill} = 0.5$ mW/cm^2 (experiment).

4.4. *Turbulence*

As the co-existence of synchrony and turbulence in the chimera state suggests, we find these states in parameter space between the fully synchronized and the turbulent states. Therefore, the chimera state is kind of an mediator between synchrony and turbulence. An experimental example of the synchronized state is shown in Fig. 3, whereas the turbulent dynamics are presented in Fig. 12.

A uniform oscillation is still present in the dynamics, but it is modulated by incoherent and aperiodic oscillations. Similar turbulent patterns have been found in Ref. 5, where also localized clusters are discussed, which are reminiscent of the subclusters presented in Fig. 10.

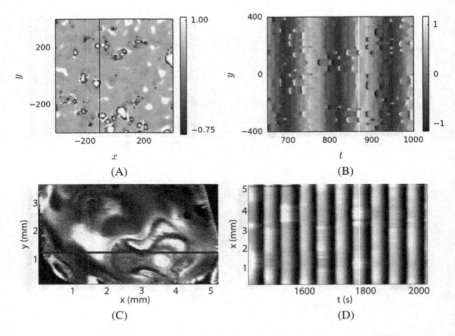

Fig. 12. Turbulent dynamics in theory (A), (B) and experiment (C), (D). Snapshots (A), (C) and one-dimensional cuts (B), (D) are shown. The whole system exhibits turbulent dynamics. Parameters read: $c_1 = 0.2$, $c_2 = -0.58$, $\nu = 0.05$, $\eta = 0.66$ (simulation) and $c_F = 50\,\text{mM}$, $\text{pH} = 3$, $R_{\text{ext}} \cdot A = 6.7\,\text{k}\Omega\text{cm}^2$, $I_{\text{ill}} = 1.0\,\text{mW/cm}^2$ (experiment).

5. Conclusions and Outlook

The oscillatory photoelectrodissolution of n-type silicon is a convenient experimental model system exhibiting a wide variety of dynamical states. Most importantly, the formation of many qualitatively different patterns is observed while the spatial average of the oscillations is simple periodic. This behavior can be well reproduced in theoretical simulations using a general normal form approach close to the Hopf bifurcation adjusted to capture the mean-field oscillation.

In the experiments, the points at the electrode surface are coupled by a linear global coupling and a nonlinear coupling both linked to the total current through the electrode surface. In the theoretical modelling this is reflected by the introduction of a nonlinear, purely global coupling. Although the details of the couplings introduced in experiment and theory are different, the results show striking similarities which led us to the conclusion that the observed phenomena are dynamical in origin and robust. Furthermore, it strongly suggests that the nonlinear coupling introduced in the experiments by the insufficient illumination is essentially global and that, e.g., the diffusion length of the minority charge carriers in the silicon plays an only minor role. An important point that might be counterintuitive at first glance, is the nature of the patterns emerging spontaneously in this purely globally coupled system. They often consist of distinct regions showing remarkably different dynamical behavior, the most astonishing example being the chimera state. Global couplings occur naturally in oscillatory systems in many fields of research.[58-61] The variety of dynamical patterns observed and their spontaneous emergence in the presented experimental system thus offers intriguing insight into the dynamical possibilities of a wide variety of systems.

Acknowledgments

The authors gratefully acknowledge financial support from the *Deutsche Forschungsgemeinschaft* (Grant no. KR1189/12-1), the *Institute for Advanced Study, Technische Universität München* funded by the German Excellence Initiative and the cluster of excellence *Nanosystems Initiative Munich (NIM)*.

References

1. J. Lechleiter, S. Girard, E. Peralta and D. Clapham, *Science* **252**(5002), 123–126 (1991), doi:10.1126/science.2011747, eprint http://www.sciencemag.org/content/252/5002/123.full.pdf.
2. A. T. Winfree, *Science* **266**(5187), 1003–1006 (1994) doi:10.1126/science.7973648, eprint http://www.sciencemag.org/content/266/5187/1003.full.pdf.
3. E. M. Cherry and F. H. Fenton, *New J. Phys.* **10**(12), 125016 (2008).
4. G. Ertl, (1991). *Science* **254**(5039), 1750–1755 (1991), doi:10.1126/science.254.5039.1750, eprint http://www.sciencemag.org/content/254/5039/1750.full.pdf.
5. V. K. Vanag, L. Yang, M. Dolnik, A. M. Zhabotinsky and I. R. Epstein, *Nature* **406**, 389–391 (2000a).
6. I. R. Epstein and K. Showalter, *J. Phys. Chem.* **100**(31), 13132–13147 (1996), doi:10.1021/jp953547m, eprint http://pubs.acs.org/doi/pdf/10.1021/jp953547m.
7. I. S. Aranson and L. Kramer, *Rev. Mod. Phys.* **74**, 99–143 (2002) doi:10.1103/RevModPhys.74.99.
8. Y. Kuramoto, *Chemical Oscillations, Waves, and Turbulence* (Dover Publications, Inc., Mineola, New York, 2003).

9. D. R. Turner, *J. Electrochem. Soc.* **105**, 402–408 (1958).
10. R. Memming and G. Schwandt, *Surf. Sci.* **4**, 109–124 (1966).
11. D. J. Blackwood, A. Borazio, R. Greef, L. M. Peter and J. Stuper, *Electrochimica Acta* **37**, 889–896 (1992).
12. K. Schönleber and K. Krischer, *ChemPhysChem* **13**, 2989–2996 (2012).
13. H. Hasegawa, S. Arimoto, J. Nanjo, H. Yamamoto and H. Ohno, *J. Electrochem. Soc.* **135**(2), 424 (1988).
14. M. Matsumura and S. R. Morrison, *J. Electroanal. Chem.* **147**, 157–166 (1983).
15. S. Cattarin, I. Frateur, M. Musiani and B. Tribollet, *J. Electrochem. Soc.* **147**, 3277–3282 (2000).
16. M. J. Eddowes, *J. Electroanal. Chem.* **280**(2), 297 (1990).
17. L. Schmidt, K. Schönleber, K. Krischer and V. García-Morales, *Chaos: An Interdisciplinary Journal of Nonlinear Science* **24**(1), 013102 (2014), doi:http://dx.doi.org/10.1063/1.4858996.
18. N. Mazouz, G. Flätgen and K. Krischer, *Phys. Rev. E* **55**, 2260–2266 (1997), doi:10.1103/PhysRevE.55.2260.
19. J. N. Chazalviel, F. Ozanam, M. Etman, F. Paolucci, L. M. Peter and J. Stumper, *J. Electroanal. Chem.* **327**(1–2), 343–349 (1992).
20. F. Paolucci, L. Peter and J. Stumper, *J. Electroanal. Chem.* **341**, 165 (1992).
21. K. Schönleber, C. Zensen, A. Heinrich and K. Krischer, *New J. Phys.* **16**, 063024 (2014).
22. I. Miethe and K. Krischer, *J. Electroanalyt. Chem.* **666**, 1 (2012).
23. V. García-Morales and K. Krischer, *Contemp. Phys.* **53**(2), 79–95 (2012), doi:10.1080/00107514.2011.642554, eprint http://www.tandfonline.com/doi/pdf/10.1080/00107514.2011.642554.
24. I. Miethe, V. García-Morales and K. Krischer, *Phys. Rev. Lett.* **102**, 194101 (2009), doi:10.1103/PhysRevLett.102.194101.
25. V. García-Morales, A. Orlov and K. Krischer, *Phys. Rev. E* **82**, 065202 (2010), doi:10.1103/PhysRevE.82.065202.
26. V. K. Vanag, A. M. Zhabotinsky and I. R. Epstein, *J. Phys. Chem. A* **104**(49), 11566–11577 (2000b).
27. A. S. Mikhailov and K. Showalter, *Phys. Rep.* **425**(23), 79–194 (2006), doi:10.1016/j.physrep.2005.11.003.
28. A. L. Lin, A. Hagberg, E. Meron and H. L. Swinney, *Phys. Rev. E* **69**, 066217 (2004), doi:10.1103/PhysRevE.69.066217.
29. P. Kaira, P. S. Bodega, C. Punckt, H. H. Rotermund and D. Krefting, *Phys. Rev. E* **77**, 046106 (2008), doi:10.1103/PhysRevE.77.046106.
30. K. Okuda, *Physica D: Nonlinear Phenomena* **63**(34), 424–436 (1993), doi:http://dx.doi.org/10.1016/0167-2789(93)90121-G.
31. H. Varela, C. Beta, A. Bonnefont and K. Krischer, *Phys. Chem. Chem. Phys.* **7**, 2429–2439 (2005), doi:10.1039/B502027A.
32. J. M. Gambaudo, *J. Diff. Eqs.* **57**, 172 (1985).
33. P. Coullet and K. Emilsson, *Physica D: Nonlinear Phenomena* **61**(14), 119–131 (1992), doi:http://dx.doi.org/10.1016/0167-2789(92)90154-F.

34. V. Petrov, Q. Ouyang and H. L. Swinney, *Nature* **388**, 655–657 (1997).
35. A. Yochelis, C. Elphick, A. Hagberg and E. Meron, *EPL (Europhysics Letters)* **69**(2), 170 (2005).
36. J. M. Conway and H. Riecke, *Phys. Rev. E* **76**, 057202 (2007), doi:10.1103/ PhysRevE.76.057202.
37. B. Marts, A. Hagberg, E. Meron and A. L. Lin, *Chaos* **16**, 037113 (2006).
38. A. Yochelis, C. Elphick, A. Hagberg and E. Meron, *Physica D: Nonlinear Phenomena* **199**(12), 201–222 (2004), doi:http://dx.doi.org/10.1016/ j.physd.2004.08.015, trends in Pattern Formation: Stability, Control and Fluctuations.
39. K. Kaneko, *Physica D: Nonlinear Phenomena* **41**(2), 137–172 (1990), doi: http://dx.doi.org/10.1016/0167-2789(90)90119-A.
40. M. R. Tinsley, N. Simbarashe and K. Showalter, *Nat. Phys.* **8**, 662–665 (2012), doi:10.1038/nphys2371.
41. D. M. Abrams and S. H. Strogatz, *Phys. Rev. Lett.* **93**, 174102 (2004), doi: 10.1103/PhysRevLett.93.174102.
42. Y. Kuramoto and D. Battogtokh, *Nonlin. Phenom. in Complex Syst.* **5**, 380–385 (2002).
43. S.-i. Shima and Y. Kuramoto, *Phys. Rev. E* **69**, 036213 (2004), doi:10.1103/ PhysRevE.69.036213.
44. E. A. Martens, C. R. Laing and S. H. Strogatz, *Phys. Rev. Lett.* **104**, 044101 (2010), doi:10.1103/PhysRevLett.104.044101.
45. G. C. Sethia, A. Sen and F. M. Atay, *Phys. Rev. Lett.* **100**, 144102 (2008), doi:10.1103/PhysRevLett.100.144102.
46. D. M. Abrams, R. Mirollo, S. H. Strogatz and D. A. Wiley, *Phys. Rev. Lett.* **101**, 084103 (2008), doi:10.1103/PhysRevLett.101.084103.
47. I. Omelchenko, Y. Maistrenko, P. Hövel and E. Schöll, *Phys. Rev. Lett.* **106**, 234102 (2011), doi:10.1103/PhysRevLett.106.234102.
48. I. Omelchenko, B. Riemenschneider, P. Hövel, Y. Maistrenko and E. Schöll, *Phys. Rev. E* **85**, 026212 (2012), doi:10.1103/PhysRevE.85.026212.
49. S. Nkomo, M. R. Tinsley and K. Showalter, *Phys. Rev. Lett.* **110**, 244102 (2013), doi:10.1103/PhysRevLett.110.244102.
50. I. Omelchenko, O. E. Omelchenko, P. Hövel and E. Schöll, *Phys. Rev. Lett.* **110**, 224101 (2013), doi:10.1103/PhysRevLett.110.224101.
51. N. C. Rattenborg, C. J. Amlaner and S. L. Lima, *Neurosc. and Biobehav. Rev.* **24**, 817–842 (2000).
52. C. G. Mathews, J. A. Lesku, S. L. Lima and C. J. Amlaner, *Ethology* **112**, 286–292 (2006).
53. T. P. Vogels, K. Rajan and L. F. Abbott, *Annu. Rev. Neurosci.* **28**, 357–376 (2005).
54. D. Barkley and L. S. Tuckerman, *Phys. Rev. Lett.* **94**, 014502 (2005), doi: 10.1103/PhysRevLett.94.014502.
55. A. M. Hagerstrom, T. E. Murphy, R. Roy, P. Hövel, I. Omelchenko and E. Schöll, *Nat. Phys.* **8**, 658–661 (2012), doi:10.1038/nphys2372.

56. E. A. Martens, S. Thutupalli, A. Fourrière and O. Hallatschek, *Proc. Natl. Acad. Sci.* **110**(26), 10563–10567 (2013), doi:10.1073/pnas.1302880110, eprint http://www.pnas.org/content/110/26/10563.full.pdf+html.
57. M. Wickramasinghe and I. Z. Kiss, *PLoS ONE* **8**(11), e80586 (2013), doi: 10.1371/journal.pone.0080586.
58. H.-G. Purwins, H. U. Bödeker and S. Amiranashvili, *Adv. Phys.* **59**, 485 (2010).
59. K. Krischer, Nonlinear dynamics in electrochemical systems, in R. C. Alkire and D. M. Kolb (eds.), *Advances in Electrochemical Science and Engineering*, **8**, 89–208 (2003).
60. S. H. Strogatz, *Physica D: Nonlinear Phenomena* **143**(14), 1–20 (2000), doi: http://dx.doi.org/10.1016/S0167-2789(00)00094-4.
61. A. Wacker and E. Schöll, *J. Appl. Phys.* **78**(12), 7352–7357 (1995), doi:http://dx.doi.org/10.1063/1.360384.

Chapter 15

OPTIMAL CONTROL OF ENTRAINMENT OF NONLINEAR OSCILLATORS WITH WEAK FEEDBACK AND FORCING

Yifei Chen and István Z. Kiss

Department of Chemistry, Saint Louis University,
3501 Laclede Ave, St Louis, MO 63103, USA

1. Introduction

Entrainment of oscillators is a fundamental concept in nonlinear science with widespread applications.[1,2] During mutual entrainment coupling between two (often slightly heterogeneous) oscillatory systems results in a state where the systems adjust their behavior to yield oscillations with a common, identical frequency. Similarly, an external periodic forcing can be regarded as a form of unidirectional coupling between two systems; during external entrainment the oscillatory system adjusts its frequency to that of the forcing signal. In biology, when the physiological synchrony between two systems (or a system and a forcing signal) breaks up dynamical diseases[3] (e.g., sleep disorders,[4] heart arrhythmia,[5] or epilepsy[6]) can occur. Therefore, there is a need for a control technique that applies an external control signal that steers the system to a desired synchronization structure.[7]

Efficient design of synchronization engineering demands a theory for optimal control. For open loop control the question is what type of waveform shall be applied to attain a certain synchronization target. For closed loop control, the feedback parameters shall be optimized to induce the required structure. Achieving optimality in these control techniques proved to be a challenge. Techniques based on accurate models[8–10] for nonlinear systems require many variables and parameters that are difficult to obtain in control application. Simple, generic models (e.g., simple phase model[11] or prototype neuron models[12]) can give only qualitative guidelines for control design.

In this chapter, we review progress on optimal control of synchronization properties of oscillators based on a simple yet accurate experiment-based phase model description.[13] The emphasis is on how to address problems related to optimal design with both closed loop and open loop approaches.

2. Synchronization Engineering with Closed Loop Feedback

A versatile closed loop feedback technique was developed based on phase model description of oscillators to steer dynamical systems to desired states.[7] The desired behavior of a population of N oscillators is engineered through imposition of a nonlinear, time-delayed feedback to a system parameter p:

$$\delta p(t) = \frac{K}{N} \sum_{k=1}^{N} h(x_k(t)) \tag{1}$$

where $\delta p(t)$ is the parameter perturbation, $x_k(t)$ is the observable variable of the kth oscillatory system, K is the feedback gain, and h is the feedback function

$$h(x) = \sum_{n=0}^{S} k_n x(t - \tau_n)^n \tag{2}$$

where S is the overall order of the feedback, and k_n and τ_n are the gain and the delay of the nth order feedback, respectively. The synchronization engineering method[7] provides a framework for obtaining the order and time delays best suited for the desired structure by describing the feedback induced structures using phase models. A population of oscillators with weak global interactions can be described[14] by

$$\frac{d\phi_i}{dt} = \omega_i + \frac{K}{N} \sum_{j=1}^{N} H(\phi_j - \phi_i) \tag{3}$$

where ϕ_i and ω_i are the phase and natural frequency of the ith oscillator, and H is the interaction function. The phase model shows that without feedback ($K = 0$) the phase of the oscillation increases linearly with a rate of ω_i. When feedback is present, the rate of phase change is slightly modified due to the coupling contribution.

The following steps should be followed for a phase model based synchronization engineering procedure:

(i) Pick a target synchronization structure, e.g., a constant phase difference between two oscillators or a structure. This structure will impose a target state for the variation of phases.

(ii) For the target synchronization structure pick a phase model that describes the system. Harmonics (sin and cos components) of H determines the behavior of the system.

(iii) Measure phase response function of a single oscillator. The phase response function (Z) is the infinitesimal phase response curve (PRC) widely used to interpret external entrainment.[14] Z is the phase advance per unit pulse perturbation as a function of the phase of the oscillator. We can obtain $Z(\phi)$ by applying thin pulses at different parts of the oscillations and measuring the phase advance. Alternate methods also exist, for example, with a feedback technique[15,16] or by integration of the spike triggered average curve.[17]

(iv) The feedback parameters to attain the desired synchronization structure through the proper interaction function H are determined with a relationship between H and Z, h, and x.[7]

$$H(\Delta\phi) = \frac{1}{2\pi} \int_0^{2\pi} Z(\phi) h(\Delta\phi + \phi) d\phi \qquad (4)$$

Therefore, a feedback loop with optimized k_n and τ_n values can be designed such that the desired H in Eq. (3) coincides with the induced H in Eq. (4). The feedback parameters can be obtained with standard optimization techniques for a single target H. In addition, there exists an approximate relationship between the feedback parameters and H: jth order feedback typically enhances the jth harmonics in the interaction function, and the feedback delay simply shifts the balance between the cosine and sinusoidal terms of the appropriate harmonics.[18]

The feedback now can be applied and it is expected that the phases of the oscillators will adjust to provide the expected synchronization structure after a transient time, which is determined by the feedback gain. A difficulty with the application of the method is the possible existence of more than one solutions to the phase model Eq. (3) with the same H. Solutions of this problem include search for H function that gives only unique, desired attractor, turning on the feedback signal only when initial conditions belong the basin of attraction of the phase model (Eq. (3)), or the application of targeting procedures to create favorable initial conditions for the application of the feedback.

The synchronization engineering technique[7] was demonstrated for a range of applications:

- Control of phase difference between two oscillators.[15] This problem was demonstrated in experiments with electrochemical oscillators[15] and with two spiking neurons.[19]
- Clustering of periodic[18,20] and chaotic[16] oscillators. The target dynamics is a predefined number of groups of oscillators; within each groups the phase of the oscillators are identical but different than those of the other groups.
- Desynchronization of oscillators.[7] When some (unknown) inherent coupling among the oscillators causes undesired synchrony, one can design a feedback to produce a completely desynchronized behavior where the phases of the oscillators are nearly uniformly distributed.

The feedback method operates in the framework of phase model approximation. Therefore, the method produces a mild, non-destructive feedback that causes only changes in the phases of the oscillators but not in the amplitude. Nonetheless, after achieving a synchronization target, the feedback signal is non-vanishing, therefore, an important question of the engineering methodology is to what extent the designed feedback is optimal to induce the desired structure. At present, there is no comprehensive approach; optimality can be considered at various stages of application of synchronization engineering.

2.1. *Optimal choice of observable and control parameters*

At the initial stage of the application, a choice for the observable and control parameter has to be made. The chosen observable should be able to reconstruct the phase of the oscillator properly. Although for most periodic oscillators any variable could suffice, the observability of phases of oscillators for chaotic oscillators often impacts synchronizability.[21] For control parameters, it would be desired that interaction function with large amplitude and higher harmonic components are present for the manipulations of the harmonic components of H. Except from such rather general arguments, the optimal choice of observable and control parameters constitutes an open problem of synchronization engineering.

2.1.1. *Optimization of interaction function H*

Many interaction functions could be possible to induce the same type of synchronization phenomena. If such situation arises, further requirements (e.g., Lyapunov stability (strongly attracting solution), uniqueness, or structural stability) can be considered in the selection of the proper H.

Such an optimal H selection was applied in the synchronization engineering problem of control of phase difference between the oscillators.[15] Many interaction functions of the form

$$H(\Delta\phi) = \sin(\Delta\phi + \Delta\tau) + R\sin(2(\Delta\phi + \Delta\tau)) \qquad (5)$$

in Eq. (3) can result in stable stationary phase difference solution between the identical oscillators in the range of $[-\pi, \pi]$ by proper choice of the major bifurcation parameter, $\Delta\tau$ (related to the delay in the feedback signal). Figure 1A shows the stationary phase difference solution of the phase model with $R = 0.1$. By changing the $\Delta\tau$ bifurcation parameter the stationary steady state solutions can be tuned in a very narrow range of $\Delta\tau$. Application of such interaction function would be impractical because small heterogeneities of the system would shift this diagram causing the phase difference between the oscillators drift quickly.

The value of R in Eq. (5) determines the nontrivial branch of the bifurcation diagram; with increasing the value of R from 0 to 0.5, the width of the nontrivial branch monotonically increases (see Fig. 1B for bifurcation diagram at $R = 0.5$). However, when $R > 0.5$, sub-critical bifurcation takes place that would create co-existence of multiple stable phase differences. Therefore, as a compromise between maximizing the width of the nontrivial branch of the bifurcation diagram and requirement for uniqueness of solution, an optimal interaction function with $R = 0.5$ was proposed for the phase difference control problem.[15]

This interaction function was found to be efficient in controlling the phase difference for electrochemical oscillators.[15] The electrochemical oscillator system of nickel dissolution in sulfuric acid produces an oscillatory

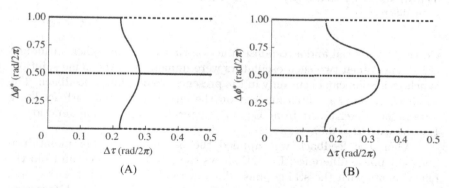

Fig. 1. Steady state solutions for phase model Eq. (3) with interaction functions used for tuning phase difference between the oscillators Eq. (5). (A) $R = 0.1$ provides narrow range of nontrivial branch. (B) Optimized interaction function with $R = 0.5$.[15]

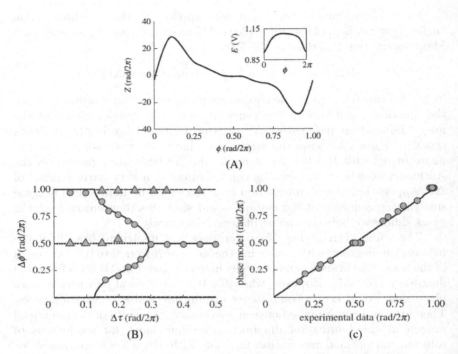

Fig. 2. Tuning phase difference between two electrochemical oscillators with closed loop feedback. (A) Response function and waveform (inset) of the electrochemical oscillator. (B) Stationary phase difference obtained with optimized quadratic feedback. Circles and triangles represent results from positive and negative overall feedback gain, respectively. Lines represent phase model predictions of the stable (solid) and unstable (dotted) stationary states. (C) Predicted versus experimentally measured stationary phase differences.[15]

electrode potential and a response function that contains higher harmonics (Fig. 2A). When two such oscillators were immersed in the same solution, synchronization can occur only in the presence of coupling or feedback. The feedback signal was superimposed on the circuit potential with feedback parameters determined from Eq. (4) to produce interaction function in Eq. (5) with $R = 0.5$.

When the feedback was applied the oscillators quickly maintained constant phase difference; Fig. 2B shows the bifurcation diagram from the experiments with the stable phase differences obtained with positive, and unstable phase differences obtained with negative feedback. The bifurcation diagram (symbols) well reproduced the expected bifurcation diagram (lines). By applying overall positive feedback gains and overall delay times that fall into the nontrivial bifurcation branch the phase differences of the oscillators can be tuned as it is shown in Fig. 2C. The experiments thus confirmed that the phase difference between the oscillators can be effectively tuned with synchronization engineering method. A similar approach was successfully implemented for engineering in-phase or anti-phase synchrony between the spiking of two patched-clamped rat hippocampal neuron cells.[19]

2.2. *Single objective optimization for weak feedback*

It is possible that preselection of a single given target interaction in Eq. (3) is not possible. For example, for desynchronization of a population of oscillators the following family of target interaction function was proposed[18,20]

$$H(\Delta\phi) = -b_1 \sin(\Delta\phi) - \sum_{k=2}^{M} \varepsilon_k \sin(k\Delta\phi) \tag{6}$$

where M is the largest harmonics considered and $\varepsilon_k > 0$ are small numbers, and $b_1 \gg \varepsilon_k$. This interaction function imposes a large negative sinusoidal term that produces phase repulsive coupling.

However, higher harmonics should also be considered in order to destroy the inadvertent stabilization of higher order cluster states. Equation (6) represents a family of interaction functions, each of which results in the desired state. Desynchronization of oscillators should be done preferably by mild, effective feedback signal. The optimization to produce weak feedback signal with a third order feedback requires setting of six parameters (k_1, k_2, k_3) and (τ_1, τ_2, τ_3) in Eq. (2). The optimization was performed in a two-step procedure:

Step 1: Since the linear feedback with small delay destabilized the one cluster state, the linear delay was set to a small value $\tau_1 = 0.01$ rad/2π. For the delay of the second order feedback a purely second order feedback was applied with τ_2 in a range of 0.01–0.12 rad/2π. A value of τ_2 was selected such that the sum of the magnitude of the first five harmonics of H with the quadratic feedback has a maximum. Such feedback maximizes the impact of the quadratic feedback. Similar approach was used to determine the third order feedback delay, but with purely third order feedback.

Step 2. With equal values of the gains (k_1, k_2, k_3) a purely linear feedback was approximately 10 times stronger than a quadratic feedback, measured by the power of the feedback signal; similarly, the quadratic feedback was approximately ten times stronger than the cubic. Therefore, in order to obtain a mild feedback that desynchronizes the system, we minimized the cost function $100k_1 + 10k_2 + k_3$ with conditions that the ε_k values of the interaction functions satisfy the following criteria

$$b_1 > 0.4, \quad \varepsilon_2 > 0.2, \quad \varepsilon_3 > 0.2, \quad \varepsilon_4 > 0.05, \quad \varepsilon_5 > 0.05.$$

The actual numbers in these criteria reflect the fact that linear feedback induces attractive second and third harmonic components that the feedback perturbation should destroy. The optimization problem for k_1, k_2, and k_3 values were solved with linear programming. Figure 3 shows that the feedback is capable of desynchronizing a population of 64 electrochemical oscillators that were strongly coupled with a common resistor. Without feedback, the system synchronizes to a one cluster state. Upon application of the feedback, desynchronization occurs; when feedback is turned off the system returns to the one cluster synchronized state. These results thus demonstrate that given a family of interaction function, a single objective cost function (related to the power of the feedback signal) can be minimized within a constraint of mild feedback signal.

2.3. Multi-objective feedback optimization

Instead of the single objective optimization used for desynchronization of a population of oscillators, a multi-objective optimization was applied for engineering cluster formation.[18,20] A phase model can produce a state, which is composed of M clusters in which each elements have identical states. Nearly balanced M cluster configurations can be obtained with the

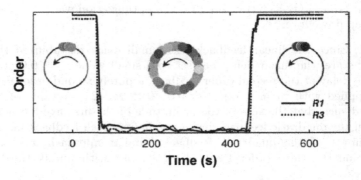

Fig. 3.   Successful desynchronization of a population of 64 coupled electrochemical oscillators. Time series of Kuramoto order parameters for one (R1) and three-cluster (R3) states before, during ($60\,\mathrm{s} < t < 440\,\mathrm{s}$), and after the optimized feedback signal was applied. Circles represent snapshot of phases of oscillators on the unit circle.[7]

following family of interaction function:

$$H(\Delta\phi) = \sum_k A_k \sin(k\Delta\phi) + \sum_k B_k \cos(k\Delta\phi).$$

(i) for any k, $B_k = 0$
(ii) for $k = M$, A_k is a large positive number
(iii) for $k \neq M$, A_k are small negative numbers.

For each harmonics k, an upper and a lower bound (LB_k and UB_k) was defined so that any interaction function within these bounds yielded stable M cluster state and all other cluster states are unstable.[22] For example, for a two-cluster state the LB values were -1.0, 0.5, -2.0, -0.8, -0.5, -0.5, -05, and the UB values -0.5, 1.0, -04, -0.3, -0.1, -0.05, -0.05 for $k = 1, \ldots, 7$, respectively. The optimization was performed by minimizing the fitness function

$$\text{Fitness} = \text{Magnitude} + \text{Penalties}$$

where the magnitude is similar to that used in desynchronization

$$magnitude = K \sum_{n=1}^{S} \frac{|k_n|}{10^n}$$

and the penalties ensure that harmonics of the interaction function are within bounds:

$$penalties = \sum_{n=1}^{7} P_n$$

$$P_n = \begin{cases} \left| B_n - \dfrac{LB_n + UB_n}{2} \right| & \text{for } B_n < LB_n \text{ or } B_n > UB_n \\ \\ 0 & \text{for } LB_n < B_n < UB_n. \end{cases}$$

After minimization of penalties the feedback was applied to a population of electrochemical oscillators. Figure 4 shows that predominantly linear, quadratic, cubic, or quartic feedback causes 1, 2, 3, or 4 cluster states, respectively. The experiments that show the multiobjective feedback optimization is capable of producing a desired cluster state.

3. Optimal Entrainment with Open Loop Feedback

Although the synchronization engineering[7] technique proved to be an efficient tool for design of synchronization structure, it requires a real-time

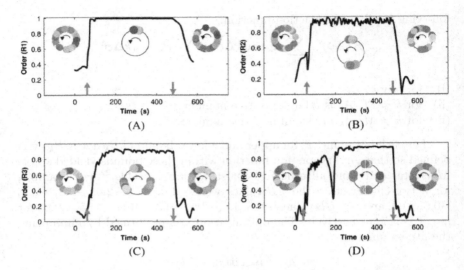

Fig. 4.   Stabilization of cluster patterns in a population of 64 electrochemical oscillators. The snapshot of phases and the time series of corresponding order parameters (R1, R2, R3, and R4) are shown for (A) one-cluster, (B) two-cluster, (C) three-cluster, and (D) four-cluster states. Arrows indicate the application and termination of the feedback.[18]

feedback control that can be difficult to implement. Open loop feedback control of oscillatory systems could provide an alternative means to tune synchronization structure. Cardiac pacemakers[5,23] and circadian entrainment by light[8,24,25] are two practical examples where open loop control has already been applied to address synchronization problems.

Most optimal entrainment open loop controls were performed with a single oscillator. Consider a Winfree phase model[2] for a single oscillator

$$\frac{d\phi}{dt} = \omega + Z(\phi)u(\theta)$$

$$\frac{d\theta}{dt} = \Omega$$

(7)

where $u(\theta)$ is the control signal coming for the external pacemaker with frequency Ω, and θ is the phase of the forcing signal. In this review we consider four optimal control applications related to frequency control of an oscillator, widening the entrainment region, fast entrainment, and classical phase control.

3.1. *Frequency control*

One of the simplest control applications is the effective modification of the frequency of an oscillator from the natural frequency ω to a target frequency Ω. Different waveform of control signals $u(\theta)$ are expected to perform differently. A simple basis of comparison is the minimum power waveform $\langle u^2(\theta) \rangle$ that can produce entrainment. ($\langle \cdot \rangle$ denotes the average by θ over its period 2π, i.e., $\langle \cdot \rangle \equiv (2\pi)^{-1} \oint d\theta$).

The constraint minimization problem was solved with a Lagrange multiplier method.[26] The result is that the optimal waveform is equivalent to Z or $-Z$ depending on the positive or negative detuning of the system. This intuitive result implies that the response function is not simply a fundamental oscillator property, but it also represents an forcing waveform that can effectively change the frequency of the oscillator.

3.2. *Widening of Arnold tongue width optimal forcing*

Efficient entrainment can be considered as wide Arnold tongue (entrainment region) in the forcing amplitude versus frequency parameter space. For example, this locking range based quantity has been widely used in phase-locked loop circuits in electronics.[27] The entrainment region for a 1:1 resonant region (where the entrainment frequency is close to the natural frequency) can be calculated from the phase difference equation by averaging Eq. (7) for one

$$\frac{d\Delta\phi}{dt} = \Delta\omega + \Gamma(\Delta\phi) \tag{8}$$

where $\Delta\phi = \theta - \phi$ is the phase difference between the system and the forcing, $\Delta\omega = \Omega - \omega$ is the frequency detuning, and $\Gamma(\Delta\phi) = \langle Z(\theta + \Delta\phi)u(\theta) \rangle$ is an interaction function (similar to H for mutual entrainment in Eq. (3)). Stationary solution (entrainment) of Eq. (8) is equivalent to $d\Delta\phi/dt = 0$; therefore the entrainment range $\Delta\omega$ is the difference between the maximum (Γ_+) and minimum (Γ_-) values of Γ. Consequently, the mathematical formulation of optimal entrainment encompasses a maximization of ($\Gamma_+ - \Gamma_-$) at fixed control power $\langle u^2(\theta) \rangle$. This problem can be solved with Lagrange multiplier method.[28] The optimal waveform can be expressed as

$$u(\theta) = Z(\theta) - Z(\theta - \theta^*) \tag{9}$$

where θ^* is a parameter obtained from a balance equation:

$$\langle Z'(\theta + \theta^*)Z(\theta) \rangle = 0. \tag{10}$$

We thus see that the optimal waveform is the difference between the shape of the response function and its θ^* shifted equivalent. (Note that Eq. (9) only describes the waveform, the amplitude of the waveform should be rescaled to match the required power.) With twice-differentiable, continuous Z there exists a trivial nonzero solution for Eq. (10) with $\theta^* = \pi$. We call this optimal waveform "generic optimal waveform" for entrainment with the shape of $Z(\phi) - Z(\phi - \pi)$. The optimal waveform Eqs. (9) and (10) have some useful properties[28]:

(i) For oscillator very close to Hopf bifurcation, the response function contains only first harmonic, therefore the optimal signal is simple sinusoidal waveform.

(ii) For weakly nonlinear oscillations with $Z(\phi) = \sin(\phi) + a\sin(2\phi)$, the higher harmonic term does not contribute to more complex waveform; the optimal waveform is still sinusoidal.

(iii) For strongly anharmonic PRC other, non-generic waveforms can exist with $\theta^* \neq \pi$. Figure 5 shows an experimental example with electrochemical oscillator.[28] The system exhibits periodic oscillations (Fig. 5A) and a response function that contains higher harmonics (Fig. 5B). Figure 5C shows the generic optimal waveform $Z(\phi)-Z(\phi - \pi)$. Figures 5D and 5E show the nontrivial shapes of two additional non-generic waveforms. The experiments measured a reduced entrainability parameter[28] that characterizes the width of entrainment per square root of waveform amplitude for various signals. The non-generic optimal waveforms outperformed the entrainability of the generic waveform, which in turn outperformed standard waveforms (sinusoidal, sawtooth, or square waves).

3.3. Fast entrainment

For many systems a design goal could be to shorten the time required for synchronization to an external forcing signal. The technique introduced for Arnold tounge widening can be expanded for fast local synchronization, where the stability of the entrained solution is maximized, i.e., $Z'(\Delta\phi)$ is minimized for at constant power constraint.[29]

The Lagrange multiplier method gives the following (unique) optimal waveform solution

$$u(\theta) = \frac{Z'(\theta)}{2\lambda} - \frac{\Delta\omega Z(\theta)}{\langle Z^2 \rangle}, \quad \lambda = -\frac{1}{2}\sqrt{\frac{\langle (Z')^2 \rangle}{P - \frac{(\Delta\omega)^2}{\langle Z^2 \rangle}}}. \tag{11}$$

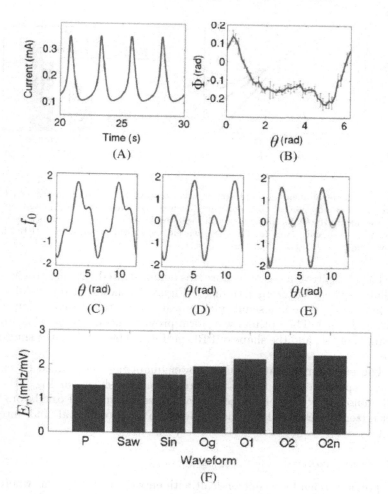

Fig. 5. Arnold tongue widening with optimal forcing of a moderately relaxational electrochemical oscillator. (A) Oscillator waveform. (B) Response function. (C)–(E) Optimal waveforms. (C) Generic optimal waveform (Og). (D) Optimal waveform 1 (O1). (E) Optimal waveform 2 (O2). Thin line: noisy optimal waveform 2 (O2n). (F) Reduced entrainabilities (width of Arnold tongue over square root of waveform power) of the tested waveforms.[28]

For zero detuning ($\Delta\omega$) the waveform is proportional to $-Z'$; therefore, for the fast entrainment problem the important quantity is $-Z'$ and not Z. For large detuning, the control must ensure frequency control as well; thus, as detuning grows, the waveform shifts into the shape of Z used for frequency control.

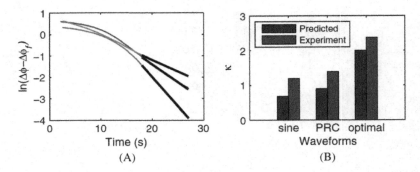

Fig. 6.   Fast entrainment of an electrochemical oscillator at zero detuning. (A) Phase difference as a function of time for sine (dashed), Z waveform (thin), and optimal $-Z'$ waveform (thick) forcing on a semi-logarithmic plot. (B) Convergence rate over square root of signal power for the three waveforms shown in panel (A). (PRC is the Z waveform.)[29]

The fast entrainment procedure was demonstrated with electrochemical oscillator system with zero detuning.[29] Figure 6A shows the phase difference as a function of time in a semilogarithmical plot for three waves (sinusoidal, PRC, and $-Z'$). The optimal waveform provides fastest decay close to the entrained phase, and the shape of PRC performed better than the sinusoidal wave.

The entrainment rates/square power quantities determined by experiments reproduced the theoretical predictions as shown in Fig. 6B. The methodology is very well suited for fast re-entrainment of oscillators that intermittently break phase locking due to environmental and internal effects.

3.4. *Phase control*

The previous open loop control dealt with entrainment problems where the initial phase of the oscillations was not known and the goals were general "dynamical" targets. However, entrainment can be considered as a classical control problem in which a signal is sought to steer the phase from initial (ϕ_i) to a final (ϕ_f) phase within a given time T_1. The mathematical problem is defined as

$$\frac{d\phi}{dt} = \omega + Z(\phi)u(t) \tag{12}$$

such that

$$\phi(0) = \phi_i, \quad \phi(T_1) = \phi_f \tag{13}$$

and the power of control signal is minimized

$$\min_{u(t)} \int_0^{T_1} u^2(t)dt. \tag{14}$$

Solution for such problems can be analytically obtained.[30] For example, when $\phi_i = 0$ and $\phi_f = 2\pi$ this problem translates into firing a neuron within a given amount of time. For $Z = \sin(\phi)$ the solution can be obtained as

$$u(t) = \frac{-\omega + \sqrt{\omega^2 - \omega\lambda_0 \sin^2 \psi(t)}}{\sin(\psi(t))}$$

where $\psi(t)$ is the optimal trajectory obtained from $\psi(0) = \phi_i$ and

$$\frac{d\psi}{dt} = \sqrt{\omega^2 - \omega\lambda_0 \sin^2 \psi(t)}$$

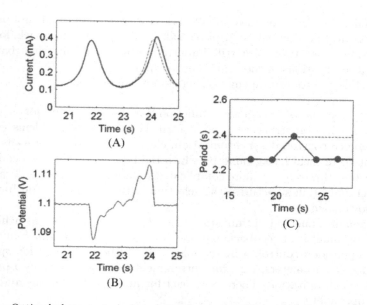

Fig. 7. Optimal phase control of an electrochemical oscillator. (A) Current versus time for the uncontrolled (dashed) and the controlled system. (B) Optimal power waveform for changing peak-to-peak period superimposed on the circuit potential $V = 1.10$ V. (C) Period of oscillations for two cycles before control, during control, and two cycles after control. Lower (red) line: Natural period $T_0 = 2.26$ s. Upper (black) line: expected controlled period $T_1 = 1.05T_0$.

where λ_0 is obtained from the following integral

$$T_1 = \int_0^{2\pi} \frac{1}{\sqrt{\omega^2 - \omega\lambda_0 \sin^2 \psi}} d\psi.$$

The control signal was derived for many common neuron model and for amplitude bounded control as well.[30]

The phase control is demonstrated with the electrochemical oscillator in Fig. 7. The system provides stable oscillations with $T_0 = 2.26\,\mathrm{s}$ without any control signal. The optimization Eqs. (12)–(14) were solved with the experimental Z function for the system and the waveform was superimposed on the circuit potential as shown in Fig. 7B. As a result of the control for one full cycle, the period was changed to $T_1 = 2.37\,\mathrm{s}$. For the next cycle the control is turned off and thus the system oscillates with the natural period (Fig. 7C).

4. Concluding Remarks

Optimal control of nonlinear oscillators is a formidable task but recent progress in phase model description aided solving weak feedback/forcing approximations. There are still important open problems that bar the efficient applications of the control procedures.

With synchronization engineering closed loop feedback:

• The selection of interaction function for phase model for a given synchronization structure is not trivial. For the given problems (phase difference control, desynchronization, cluster formation) there was sufficient theoretical background by which the interaction function could be selected. However, for more complicated functions (e.g., for time varying synchronization structures) the selection criteria of interaction function is not known.

• Although numerical optimization of feedback parameters was achieved, the optimality of feedback parameters was not confirmed. Similar to the open loop control, a methodology should be developed for optimal feedback parameters, e.g., for entrainment of wide frequency range or fast synchronization. There is a need for analytical solutions and tests for optimality.

With open loop control:

• The methodologies were developed primarialy for 1:1 resoncance control. Although the phase model description can be easily adopted for m:n

entrainment, the properties of the waveforms for higher order entrainements are largely unknown.

- The open loop control techniques have been tested with a single oscillator. Extensions to optimal entrainment of multiple oscillators shall be made. For oscillator populations multiple approaches are possible with global forcing, localized "pinning" control,[31] or combinations.

For both open and closed loop control methodology the theories should be extended to noisy oscillators, to networks, and to large control signals that are beyond the phase description validity. Although some initial attempts have been made for the two former examples, (e.g., synchronization engineering of chaotic oscillators[16] or optimal entrainment of neural ensembles[32]) the strong signal approximation requires extension of phase model description to large signals. Such theoretical framework exists,[33] but its use in optimal control is topic for future investigation.

Acknowledgments

This material is based upon work supported by the National Science Foundation under Grant No. CHE-0955555. Acknowledgment is made to the President's Research Fund of Saint Louis University for support of this research.

References

1. A. S. Pikovsky, M. G. Rosenblum and J. Kurths, *Synchronization — A Universal Concept in Nonlinear Sciences* (Cambridge University Press, Cambridge, England, 2001).
2. A. T. Winfree, *The Geometry of Biological Time* (Springer-Verlag, New York, 1980).
3. J. Bélair, L. Glass, U. An Der Heiden and J. Milton, *Chaos* **5**(1), 1–7 (1995).
4. D. A. Golombek and R. E. Rosenstein, *Phys. Rev.* **90**(3), 1063–1102 (2010).
5. L. Glass, *Chaos* **1**(1), 13–19 (1991).
6. F. L. da Silva, W. Blanes, S. N. Kalitzin, J. Parra, P. Suffczynski and D. N. Velis, *Epilepsia* **44**, 72–83 (2003).
7. I. Z. Kiss, C. G. Rusin, H. Kori and J. L. Hudson, *Science* **316**(5833), 1886–1889 (2007).
8. N. Bagheri, J. Stelling and F. J. Doyle, *PLoS Comput. Biol.* **4**(7), e1000104 (2008).
9. S. J. Schiff, *Philos. Trans. R. Soc. A* **368**(1918), 2269–2308 (2010).
10. D. Lebiedz, *Int. J. Mod. Phys. B* **19**(25), 3763–3798 (2005).
11. P. A. Tass, *Phase Resetting in Medicine and Biology. Stochastic Modeling and Data Analysis* (Springer, Berlin, 1999).

12. M. Rosenblum and A. Pikovsky, *Phys. Rev. E* **70**(4), 041904 (2004).
13. I. Z. Kiss, Y. M. Zhai and J. L. Hudson, *Phys. Rev. Lett.* **94**(24), 248301 (2005).
14. Y. Kuramoto, *Chemical Oscillations, Waves and Turbulence* (Springer, Berlin, 1984).
15. C. G. Rusin, H. Kori, I. Z. Kiss and J. L. Hudson, *Philos. Trans. R. Soc. A* **368**(1918), 2189–2204 (2010).
16. C. G. Rusin, I. Tokuda, I. Z. Kiss and J. L. Hudson, *Chem. Int. Edit.* **50**(43), 10212–10215 (2011).
17. G. B. Ermentrout, R. F. Galán and N. N. Urban, *Phys. Rev. Lett.* **99**(24), 248103 (2007).
18. H. Kori, C. G. Rusin, I. Z. Kiss and J. L. Hudson, *Chaos* **18**(2), 026111 (2008).
19. C. Rusin, S. Johnson, J. Kapur and J. Hudson, *Phys. Rev. E* **84**(6), 066202 (2011).
20. C. G. Rusin, I. Z. Kiss, H. Kori and J. L. Hudson, *Ind. Eng. Chem. Res.* **48**(21), 9416–9422 (2009).
21. C. Letellier and L. A. Aguirre, *Phys. Rev. E* **82**(1), 016204 (2010).
22. K. Okuda, *Physica D* **63**(3–4), 424–436 (1993).
23. L. Enochson, J. Sandstedt, H. Strandberg, C. Emanuelsson, A. Ornberg, A. Lindahl and C. Karlsson, *J. Electrocard.* **45**(3), 305–311 (2012).
24. D. A. Dean, D. B. Forger and E. B. Klerman, *PLoS Comput. Biol.* **5**(6), e1000418 (2009).
25. A. E. Granada and H. Herzel, *PLoS One* **4**(9), e7057 (2009).
26. J. Moehlis, E. Shea-Brown and H. Rabitz, *J. Comput. Nonlin. Dyn.* **1**(4), 358–367 (2006).
27. R. E. Best, *Phase-Locked Loops: Design, Simulation, and Applications* (McGraw-Hill, New York, 1997).
28. T. Harada, H.-A. Tanaka, M. J. Hankins and I. Z. Kiss, *Phys. Rev. Lett.* **105**(8), 088301 (2010).
29. A. Zlotnik, Y. Chen, I. Z. Kiss, H. A. Tanaka and J. S. Li, *Phys. Rev. Lett.* **111**(2), 024102 (2013).
30. I. Dasanayake and J. S. Li, *Phys. Rev. E* **83**(6), 061916 (2011).
31. L. Turci and E. Macau, *Phys. Rev. E* **84**(1), 011120 (2011).
32. A. Zlotnik and J. S. Li, *J. Neural Eng.* **9**(4), 046015 (2012).
33. W. Kurebayashi, S. Shirasaka and H. Nakao, *Phys. Rev. Lett.* **111**(21), 214101 (2013).

INDEX

Printed in the United States
By Bookmasters